SHOCKING BODIES

SHOCKING BODIES

BODIES

LIFE, DEATH & ELECTRICITY
IN VICTORIAN ENGLAND

IWAN RHYS MORUS

The
History
Press

For my wife Mandy and son Gwilym Dafydd

Cover illustrations: front: French physicist Arsene D'Arsonval demonstrating the use of the human body as a conduit for the transmission of electric power, *Electrical Review*, 1894; back: lightning in my hands, stock.exchng.

First published 2011

The History Press
The Mill, Brimscombe Port
Stroud, Gloucestershire, GL5 2QG
www.thehistorypress.co.uk

British Library Cataloguing in Publication Data.
A catalogue record for this book is available from the British Library.

ISBN 978 0 7524 5800 7

Typesetting and origination by The History Press
Printed in Great Britain

CONTENTS

	Acknowledgements	6
	Prologue	7
PART I	**TOM WEEMS**	**11**
one	Tom Weems' Body	11
two	Galvanising Britain	20
three	Galvanic Fashions	29
four	Body & Soul	39
five	Dissecting Tom Weems	49
PART II	**ADA LOVELACE**	**58**
six	Knowing Ada	58
seven	Electric Universe	70
eight	Galvanic Medicine	81
nine	Ada's Laboratory	92
PART III	**CONSTANCE PHIPPS**	**102**
ten	Lady Constance's Pain	102
eleven	Electric Frontier	112
twelve	Machinery of the Body	124
PART IV	**MR JEFFERY**	**136**
thirteen	Introducing Mr Jeffery	136
fourteen	Electric Entrepreneurs	145
fifteen	Measure for Measure	159
sixteen	Harness on Trial	172
PART V	**BACK TO THE FUTURE**	**183**
seventeen	Back to the Future	183
	Notes	186
	Bibliography	206
	Index	217

ACKNOWLEDGEMENTS

In between other projects I have been working on this book for quite a long time. During that time conversations with a great many people have helped deepen my understanding of the history of Victorian science and culture. I am particularly grateful to Will Ashworth, Graeme Gooday, Jeff Hughes, Rob Iliffe, Bernard Lightman, Jim Moore, Richard Noakes, Simon Schaffer, Anne and Jim Secord and Andy Warwick. Some of the research was carried out with funding from the Wellcome Trust and I am grateful to the Wellcome trustees for their generosity. Much of the final writing took place during a year's leave, for which I thank Aberystwyth University and the AHRC. I also thank my editors, Lindsey Smith and Christine McMorris of The History Press, for their work on this book. The book is dedicated to my wonderful wife Mandy. It is also dedicated to our son Gwilym Dafydd, who since his arrival half way through that year of writing has been doing his level best to distract his father from finishing the job.

PROLOGUE

The Victorian age was full of shocking (and shocked) bodies. Throughout the nineteenth century, electricity was the stuff of life – and death. When people thought about their own bodies – how they worked, what the relationship was between body and soul, how the relationship between the sexes worked, or ought to work, even the politics of individual rights and obligations – they turned to electricity as a way of making sense of difficult questions. So electricity helped the Victorians to understand their own bodies – but their bodies helped the Victorians understand electricity too. We do not usually think about physics and bodies in the same breath, or at least not the messy, shambolic, everyday bodies we carry around with us. The bodies of physics, if it has any, are abstract, heavenly, ideal. This means that when we write histories of physics they tend to be disembodied too. Even when they are histories of great men (and they often are) they are histories that concentrate on minds, not bodies.

So what would a history of physics that took bodies seriously look like? This book is an effort to find out. In it, I try to find out how physics affected people's bodies – and how their bodies affected physics – throughout the nineteenth century in Britain. I try to show that Victorian physics was indeed corporeal and that by looking at the bodies of physics we can get quite a different idea of where science belonged in nineteenth-century culture, why it mattered, and to whom. I try to show that we cannot really understand physics without taking bodies into account, but also that we cannot really understand Victorian bodies without paying attention to the physics going on around and with them.

Looking at things from this sort of perspective allows the historian to try writing the history of Victorian physics from below. In just the same way that economic, social or political historians now tend to avoid history written from the perspective of the powerful, and try to recover the aspirations and experiences of the crowd, I think we need to look at nineteenth-century physics with a different kind of eye. I have chosen four bodies, scattered across the century, and I want to try and understand what physics – and electricity in particular – did to and with those bodies. The bodies themselves are windows into the culture of Victorian electricity seen from an unusual angle. They take us from the political turmoil of the nineteenth century's opening decades to the frenzied consumerism of the century's end. By looking at these bodies and their fates I want to try to recover something of how electricity came to permeate nineteenth-century and Victorian culture, what it represented to those that encountered it and how it changed their lives. This means looking for and finding Victorian electricity in some sometimes unexpected places. Electricity throughout the nineteenth century was forever getting away from its producers, however mightily they laboured to keep it under control.

If there is a single thread that runs through all four sections of the book, it is the emphasis on electricity as a science of wonder. One of the reasons that looking at electricity through the lens of individual bodies is so illuminating is that electricity really was a very corporeal science, embedded and embodied in spectacular performances of all kinds throughout the Victorian age. Electricity was all about show for the Victorians and their immediate predecessors. Electricity was striking because it was spectacular. Electricity was thrilling and seductive. Nineteenth-century encounters with electricity could be intimate, titillating and flamboyant. In many ways electricity was dangerous exactly because it seemed so corporeal. You could feel electricity shuddering through your body or see it lighting up the sky. If no one agreed what electricity really was, everyone agreed that it was powerful stuff. This could make materialist messages even more seductive and leave some men of science struggling to control their electric creations. Electricity and electricians did not attract their audiences by appealing to their reason, they appealed by attracting their senses instead. Electricity simply was the science of showmanship to most Victorians, and those who thought otherwise had to struggle very hard indeed to disembody this most vital of fluids.

The book's first body belongs to Tom Weems, an illiterate labourer who does not, I am afraid, survive the chapter's opening pages. Weems' body provides a way of raising the lid off the bubbling cauldron that was early nineteenth-century electrical culture. Electricity during the years of war with revolutionary and Napoleonic France was dangerous stuff – and it was its very corporeality that its many enemies regarded as its most dangerous feature. Several decades away from being a Victorian himself, Weems' body belongs in this primarily Victorian survey simply because it is impossible to understand Victorian electrical bodies other than against the backdrop of the century's early decades. Later generations of men of science would set their faces firmly in stubborn opposition to what seemed to them the excessively sensationalist science of their Enlightenment predecessors. Their electricity would need to be carefully sanitised, and we can see this process starting in Sir Humphry Davy's reaction to some of the experiments described in this section. But the events, arguments and experiments surrounding Tom Weems' body also show how difficult it was to get away from sensationalist science. Electricity remained firmly corporeal, which meant that the battle would need to be fought all over again by the Victorians.

The next body belongs to Ada Lovelace, Lord Byron's only daughter and 'Soul of my thought', as he put it in *Childe Harold's Pilgrimage*. This might seem an unlikely location for a battleground over electricity's future, but that is exactly what Ada's body became, for a while at least. By the early years of Victoria's reign, electricity remained for many of its practitioners and its audiences a very bodily business. Electricity was in many ways tangible – a mysterious fluid that might be invisible and unknown but whose effects were directly felt; intimately corporeal and sensationally spectacular. This was going to be one of electricity's paradoxes throughout the Victorian age. It was invisible, ethereal and no one really knew what it was, but at the same time its effects were unambiguously physical and thrillingly visual. Indeed it was just this quality that made electricity such a useful and ambiguous cultural resource. Whilst natural philosophers like Michael Faraday laboured hard to make electricity disembodied and reputable, Ada Lovelace's example demonstrates just how difficult that task would be. Ada's view of electricity as a corporeal agent, and her understanding of electrical science as a route towards bodily self-knowledge, was radical and transgressive for many reasons – but most importantly because of who she was. She was trying to frame a feminine view of electricity at odds with the perspective of scientific gentlemen.

Twenty years or so into Victoria's reign electricity – as the third body shows – was still stubbornly corporeal. This third body belongs to Constance Phipps, the largely forgotten daughter of politician and diplomat the second Marquess of Normanby. Her story and her short life give an insight into electricity and its operations in the mid-Victorian world of female invalidism. But it also provides some insights into the culture of electricity at its most titillating. Electricity wielded by self-styled authorities was a powerful tool to bring awkward bodies to heel. As electricity's tentacles started encircling the globe, the human body itself provided new ways of talking about the developing telegraphic world. The telegraph was 'the nervous system of Britain', after all. At the same time, the telegraph provided a whole new electrical vocabulary for talking about nervous order and disorder. Electrical showmanship was also at its most spectacular during this mid-Victorian heyday, and electricity provided a feast for the senses as well as a set of tools for keeping sensation carefully disciplined. Increasingly, electricity represented and promised the future to the Victorians. The electrical future was a world of spectacle and power in which electricity was both intimately corporeal and universal in its possibilities. For the young Constance Phipps, though, the final place of electricity was the sickroom.

The final body – belonging to the otherwise utterly anonymous and invisible Mr Jeffery – demonstrates how corporeal electricity had thoroughly become part of late Victorian consumerism by the 1890s. The future was for sale in the pages of middlebrow magazines and popular journals aimed at a new, literate and confident readership. Electricity – and the electricity of the body in particular – was central to these late Victorian imagined futures. It was also for sale. Electrical health had become a commodity and the back pages of those popular weeklies were crammed with advertisements extolling the virtues of electrical cures. Jeffery was far from the only customer attracted to these nostrums, but he was the only one (as far as we know) whose disputed cure ended up in court. The Victorian age's final two decades also became the time of scientific measurement. The new breed of specialist physicists prided themselves on their culture of precision. As we shall see, Jeffery's alleged electrical cure generated a culture clash in the courtroom between the promoters of this new discipline of accurate measurement and the vendors of the body electric. The 1890s also ushered in another new technology of bodily electricity, with the inauguration of the electric chair. Electricity now really did seem to offer both life and death.

Throughout the Victorian age, electricity's links to the body, and its role as a conduit for bodily sensation, were on show in a whole range of different places – some predictable, others rather less so. Shocking bodies could be found in dissecting rooms, in anatomical theatres, in sickrooms and hospitals. Electricity's connections to the body were on show in popular galleries of science and on city streets. The body itself became a laboratory for investigating the mysterious fluid. Even as the legal showdown between the salesmen of the healthy electrical body and the new arbiters of electrical propriety played itself out in the courtroom, the places where electrical sensations could be experienced were multiplying. Electricity's intimate connection to the body was proving extremely hard to sever. Therefore, if we want to understand how electricity was experienced by most Victorians who encountered it, and how it infiltrated Victorian culture, then we can only do so by understanding its bodily connections. Electrical culture and electrical ways of thinking are inherited by us from the Victorians; as such, understanding shocking bodies is an important part of making sense of our own contemporary relationship with electricity as well. Many of the connections that were forged then are still with us now.

Victorians thought that electricity symbolised everything that was positive about their century of progress. It was the ultimate symbol of man's ability to dominate and control the powers of nature; with electricity it seemed that you could do almost anything. During the Victorian age electricity was even seen to replace God as the ultimate tool for explaining humanity's place in the universe. For some, saying life was electrical was much the same thing as saying there was no such thing as a soul, and therefore no such thing as God. Later on in the century, the slogan was an invitation to buy new commodities like electric belts or corsets that could revitalise a flagging body. We are still intimately aware of electricity's relationship with our bodies – its possibilities and dangers. Margaret Thatcher took electric baths and Cherie Blair wears magnetic crystals to enhance her aura. This awareness is why we worry about living too close to overhead power lines or about mobile phone masts built next to our children's playgrounds. The Victorians' faith in electrical progress and their fascination with bodily electricity tells us some important and surprising things about their culture. Our contemporary worries about electricity's impact on our own bodies should tell us some important things about our culture too. And if we want to understand our fears, we need to understand their cultural history.

PART I

TOM WEEMS

ONE

TOM WEEMS' BODY

Trudging across the featureless Cambridgeshire landscape on a summer's day in 1819, Tom Weems could hardly have expected ever to take part in such a peculiar scientific ceremony. Scientific ideas and activities of any kind would almost certainly have been completely alien to an ordinary working man like him. Natural philosophy (the study of nature, as opposed to moral philosophy, the study of man) was very much the business of social elites at the beginning of the nineteenth century. A taste for polite knowledge was the mark of a cultured and sophisticated gentleman (or, more rarely, gentlewoman). They were the ones who flocked to public lectures on the latest discoveries, read about them in newspapers and in the pages of popular journals, or discussed them in clubs and in fashionable salons.[1] Any talk about the separate cultures of art and science would have seemed very strange indeed to an early nineteenth-century gentleman. Familiarity with both was taken for granted. Science was a source of genteel entertainment; experiments were expected to be sensational. They were meant to appeal to the senses as much as to the intellect. The latest scientific discoveries provided topics for cultured conversation in much the same way as did the titbits in the *Gentleman's Magazine*.

Indeed natural philosophy – and the new science of galvanism in particular – preoccupied many political radicals at the turn of the century as well. For them, science, by revealing the true character of nature's laws, would reveal the extent to which England's laws and government fell far short of nature's requirements. The social order, said the radicals, was meant to mirror the order of nature with its checks and balances to keep everything on an even keel. The laws that governed nature were meant to govern society as well. If society seemed out of step with nature, then only decisive political action could restore the balance. Such political radicals though, however much they might declaim about the rights of the ordinary working man, still lived in a very different world from that of common labourers like Tom Weems.[2] Men like John Thelwall or William Godwin were fascinated

by the latest discoveries in galvanism – the science of electricity and the body –
because they thought such discoveries demonstrated that the soul was material and
that there was no need for God. It is perfectly possible that Tom Weems, on the
other hand, had never so much as heard of electricity, until he murdered his wife.

Weems strangled his young wife, Mary Ann, on Friday 7 May 1819 as they walked
across the Cambridgeshire countryside from Godmanchester, near Huntingdon,
on their way towards London. It might look to us at first glance like just another
sordid early nineteenth-century crime of passion – grist to the mill for hack writers
grinding out cheap and lurid gothic melodramas – but its consequences led Tom
Weems on a journey into the very heart of science. Science of any kind was a long
way removed from the everyday experiences of an ordinary country working man
like Tom Weems for a number of reasons. For a rural worker at the beginning of the
nineteenth century, the main preoccupation was making sure that there was money
enough to find something to eat and a roof for the night. Weems almost certainly
could not read or write – and would have had little use for such accomplishments
even if he had the rudimentary skills. His life would have been almost entirely taken
up by sheer hard, gruelling labour for very little pay. There was certainly little spare
time available to ponder the relationship of body and soul and its implications for
the state of society.[3] His concerns were of a more immediate and urgent kind.

The Cambridgeshire countryside through which Tom and Mary Ann travelled
that Friday morning was going through its own momentous changes. A series of
enclosure acts during the first two decades of the century had led to the closing
off of common land, cutting off many of the rural poor from a vital resource. As an
overwhelmingly agricultural county, Cambridgeshire had done quite well out of
the Napoleonic Wars. High prices had filled the coffers of the landowning gentry
and led to an increase in the amount of land being farmed. But the end of the war
brought a catastrophic drop in prices and an agricultural depression. Tempers flared
as landowners returned to the tried and tested methods of squeezing the tenantry
when their own belts looked like they needed tightening.[4] There was a major riot
in the fenland town of Littleport, between Cambridge and Ely, on 18 June 1816
which spread to Ely itself a day later. Troops were called in and blood was spilt in
the ensuing mayhem before order was eventually restored at the point of a bayo-
net.[5] Rumbling discontent was to continue well into the 1820s. Five of the rioters
were hung and many transported. One of the judges on that occasion, Mr Justice
Burrough, was to be back in Cambridge again three years later, with Tom Weems in
the dock before him.

The town of Cambridge itself, where Weems was dragged before the
Cambridgeshire Assizes on Monday 3 August, was only slowly undergoing its
own transformation during the first quarter of the nineteenth century. Up until
the end of the eighteenth century, Cambridge had changed little, if at all, from
the claustrophobic market town on the edge of the Fens that it had been when
the university arrived almost half a millennium earlier. It still retained its maze of

dirty, narrow and badly-lit streets, along with housing that appeared in a state of permanent semi-collapse. The enclosure acts eventually provided space to expand beyond the medieval boundaries, and the first decades of the nineteenth century saw some belated attempts at civic improvement.[6] Cambridge was famous then, as it is now, for its university. The university colleges were the town's most prominent features; dominating the town's skyline, they owned most of the land and property too. This meant, of course, that any improvement could only take place with the consent of the colleges who, by and large, were more inclined to consult their own comfort and convenience than that of the townspeople. They remained bastions of conservatism and ingrained privilege.[7]

The town's political culture – as well as its geography – remained firmly under the thumb of the university, which jealously guarded the ancient rights accorded it by charter. The result was that the town, like the university (at least the majority of college fellows), was overwhelmingly Tory. Only a tiny proportion of the towns-men were resident freemen with the right to vote – and those not in the university's pocket belonged to the dukes of Rutland instead. Town and gown resentment sim-mered. One of the reasons the town corporation opposed the provision of street lighting for so long was that they feared better visibility would simply make it easier for potential assailants to see each other in the dark of night. There was deep animosity towards corporation and university alike amongst the disenfranchised citizenry, along with a ready ear for radical agitation. All this made for a dangerous combination of barely repressed anger in town and county, matched by a twitchy belligerence on the side of the authorities. This was not a good time to be on the wrong side of the law, for any reason. The Littleport riots had demonstrated how quickly and how easily things could get out of hand, and how savage those charged with maintaining order could be in their efforts to do so.

Tom and Mary Ann's tragedy had begun about a year or so earlier, when they were married in the parish church of Goddington in Bedfordshire.[8] Things had started badly. Weems had been forced into the marriage by the parish officials after Mary Ann claimed that she had been made pregnant by him. They separated imme-diately after the ceremony and appear never to have lived together. Matters were not, presumably, much helped by the eventual revelation that Mary Ann had never in fact been pregnant at all. Weems left the district and found himself work as a mill-hand in Edmonton – then a village in Middlesex – whilst Mary Ann returned to her family in Godmanchester. About a year after the shotgun wedding Weems set out for Godmanchester, to all appearances in an effort to persuade his recalcitrant bride to return with him to Edmonton. Mary Ann seems to have agreed cheerfully enough to her husband's proposition, as they set out early on the morning of Friday 7 May on the long walk towards London. They were spotted in a field near the little village of Wendy by a woman named Susannah Bird, who was on her way to the market town of Royston. On her way back towards Wendy that evening, she was sur-prised to see Weems on his own on the road. She approached him, asking what had

happened to the young women she had seen with him that morning. Her suspicions were raised by Weems' evasive manner, and Susannah sounded the alarm.
Mary Ann's body was found in the field where she had last seen them together. She
had been apparently strangled with one of her own garters.

With hue and cry raised against him, it was not long before Weems was brought
in. A coroner's jury was promptly sworn in and sat until the early hours of the
following Saturday morning before bringing in a verdict of wilful murder against
Tom Weems. The coroner committed Weems and he was duly hauled off to the
county gaol in Cambridge to await his fate with the coming of the biannual
Cambridgeshire Assizes at the beginning of August. Mary Ann's remains were
returned to Godmanchester. Her body was put on show in the window of a local
public house to assuage the curiosity of her former neighbours and ghoulish stran-
gers alike. She was buried in the local churchyard, although her ghost is still reputed
to haunt the tavern where her body was exhibited. The stone that was put up over
her grave carried a stern warning to its readers:

> As a Warning to the Young of both Sexes This Stone is erected by public Subscription
> over the remains of MARY ANN WEEMS who at an early age became acquainted
> with THOMAS WEEMS formerly of this Parish this connexion terminating in a
> compulsory marriage occasioned him soon to desert her and wishing to be Married
> to another Woman he filled up the measure of his iniquity by resolving to murder
> his Wife which he barbarously perpetrated at Wendy on their Journey to London
> toward which place he had induced her to go under the mask of reconciliation May
> the 7th 1819.[9]

The twice-yearly visits of the Assize courts were important dates in the social calendar, for Cambridge as for any other provincial county town on their circuit.
They were highly ceremonial and public affairs, often with the presiding judge
and court officials parading through the town streets – all pomp and circumstance
– accompanied by civic dignitaries in the full panoply of their positions. This was
a theatrical age for which the public trappings of power were a serious business.[10]
Not only were the Assizes themselves an important social gathering, they were the
occasion for other jollifications as well. Assizes provided an excuse for balls and
soirées, where town and country toffs would let their hair down and party together.
There were opportunities for meeting up with old acquaintances and for making
new ones. The well-to-do could catch up with the latest fashion and keep abreast
of the most recent political news and gossip. On an occasion like this there might
even be talk about the latest scientific controversies. For Weems, sweating it out in
Cambridge gaol, the Assizes were a less pleasant prospect. With the Coroner's verdict against him already, he can have had little doubt of the fate that waited for him.

Assizes courts followed their own particular practice. Indictments were, as a rule,
read out twice. Once in private before a Grand Jury, and then again in public before

the Petty Jury if they considered the charge worth pursuing. The judge played an active part in the proceedings, intervening to cross-examine witnesses and direct court officials, and they were rarely long, drawn out affairs. The judge on this occasion, Mr Justice Burrough, had plenty of experience to guide him through the process. Celebrating his seventieth birthday that August, he had been one of the hanging judges at the trial of the Littleport rioters a few years previously.[11] He had suitably tough views on sentencing, even by early nineteenth-century standards. From 1800 the death penalty was increasingly commuted to transportation – except in cases of wilful murder. Mr Justice Burrough, however, was not much given to such leniency. His reply only a few years later to Lord Palmerston – when that noble lord had the temerity to suggest that justice might occasionally be tempered with mercy – was telling: 'My Rule is that where a man is convicted of a Capital Offence attended with Circumstances of Wanton Cruelty, never to extend favour to the convict.'[12] He was not about to do so this time either.

Predictably enough, the trial did not detain Mr Justice Burrough for very long.[13] John Beck testified to having carried the prisoner in his chaise for part of his journey, and that Weems had told him he wanted to get rid of Mary Ann so he could marry another woman in Edmonton. Susannah Bird gave her evidence of having seen Weems and Mary Ann together on the road and, with others, described the gruesome finding of the body. Constable Jackson gave his account of Weems' apprehension. Maria Woodward was then called to describe how she had become engaged to Weems at Edmonton, having been led by him to believe he was a single man, and informed the court how Weems had told her he was going home to Godmanchester to get some money before returning to Edmonton to marry her. Finally – and damningly – Mr Orridge, the Cambridge gaoler, described how Weems had confessed the crime to his father and sister when they visited him and described the exact method of the murder: 'I grasped my hands round her throat, pressed her windpipe with the thumbs, and exclaimed "Now I'll be the death of you", and held her in that position for about 5 minutes', before tying one of her garters around her neck (to make sure, presumably) and hiding her body in a ditch.[14] It took the jury just five minutes to find him guilty, too.

Mr Justice Burrough duly 'passed the dreadful sentence of the law, and ordered him for execution on Friday the 6th instant and his body to be delivered to the surgeons for dissection'.[15] A sentence of death is certainly what Weems would have expected to receive, having been found guilty of murder. The further penalty of post-mortem dissection would, however, have been an additional blow. In the early nineteenth century dissection was the final insult that the law could inflict upon the bodies of executed criminals, and was quite explicitly intended as an additional expression of judicial outrage. It extended the physical punishment that the law could impose beyond death itself. In a culture where many took the idea of physical resurrection quite literally, it was a sentence that might seem to remove even the last hope of salvation from the condemned, involving as it did the complete disassembly of the body. Executions like these provided, in fact, the only legal source of bodies for anatomical dissection available in early nineteenth-century England, as in much of the rest of Europe. The dissections were public and often highly ceremonial affairs, with local bigwigs as well as students and medical men coming along to see medicine doing its bit in maintaining the authority and dignity of the state.[16]

Tom Weems had his day the following Friday. A few minutes after twelve o'clock, he was hung from the gateway of Cambridge's county gaol in front of a large and enthusiastic crowd. His behaviour as he was led to the scaffold was 'consistent with his awful situation, and he met his ignominious fate with firmness'.[17] His body was left hanging for an hour before being taken down. His corpse was then bundled on to a cart and transported down Castle Hill from the gaol, through the town and to the Chemical Lecture Room in the Botanical Garden. It was accompanied by an entourage of sheriff's officers and constables. This was not just a matter of ceremony — friends of the executed often tried to steal the body on such occasions in an effort to prevent the ultimate horror of public dissection. No one came to Weems' rescue though, and he was soon lying on a slab in the lecture room, waiting for the anatomist's attentions. On this occasion there was one departure from the usual course of events at such happenings. Weems' remains were first to be examined by James Cumming, Professor of Chemistry at Cambridge University, using a 'powerful galvanic battery'. The room was packed for the occasion, with 'nearly all the medical gentlemen in Cambridge', as well as 'several of the most respectable inhabitants of the town and county'.[18]

They had come as witnesses to a very unusual experiment. Weems' body was going to be galvanised in an attempt to see if it was possible to return him to life. Cumming set about his task with systematic zeal. First of all:

> wire was applied to a small incision in the skin of the neck over the par vagum, and the other to one made between the 6th and 7th rib; when at each disturbance of the battery, the chest was disturbed in a manner similar to a slight shuddering from cold; the period of the shuddering corresponding with the number of plates struck by the operator in the last trough.

Afterwards, the:

> par vagum was laid bare, and one of the wires passed under it; the other was placed in contact with the diaphragm, through an incision made deeper than the last between the 6th and 7th ribs. The contractions were evidently stronger than in the last experiment, and to all appearances confined to the same set of muscles — Not the smallest action of the diaphragm was perceptible.[19]

Cumming was trying to make Weems' corpse breathe again by exciting the par vagum nerve that ran from the brain to the heart, lungs and digestive organs. The idea was that electricity would replace the nervous fluid (the vis nervosa) that regu-lated a living body.

Cumming then moved on to galvanise other parts of the body:

wire was placed under the supra-orbitary nerve, the other remaining under the par vagum; at each discharge of the battery, it produced considerable action of the mus-cles of the face, and more particularly on the side of the face to which it was applied, though not expressive of any of the mental affections of life; it might more properly be called a convulsive twitching.

Moving on to other parts of the man's anatomy, a wire was:

passed under the ulnar nerve at its seperation from the axillary plexus, the electric circuit was completed by bringing the other in contact with the radial nerve at the wrist. The flexor muscles of the arm and hand were thrown into strong action, the arm being drawn up, and the fist closing with considerable force.

This was more like what the experimenters were looking for. Finally, the wires from the battery were placed:

in contact with the spinal marrow between the 3rd and 4th cervical vertebrae and the tibial nerve, in its passage behind the inner ankle. A more extensive, though less vigorous effect followed this exhibition of the galvanic influence than in any of the above-mentioned experiments; most of the muscles of the trunk and extremities answered feebly the discharges of the battery.[20]

Cumming was a relatively fresh incumbent of Cambridge's chair of chemistry, having succeeded Smithson Tennant to the professorship following his unexpected death in 1815, only a couple of years after being awarded the chair. Cambridge chemistry had a slightly longer history, however. There had been an established chair of chemistry since 1702 and the Chemical Lecture Room in the Botanic Garden – where the experiments on Weems' corpse took place – had been there since 1786. The galvanic battery that Cumming used on this occasion had been made especially for Tennant. Cumming himself was, nevertheless, making some-thing of a name for himself as a galvanic experimenter and must have relished the chance Weems provided him. It was a golden opportunity to make his mark as a daring and controversial experimenter. Galvanism was a political, theological and intellectual minefield – and Cumming was no doubt hoping that he would be able to use his Cantabrigian authority to stamp out some of the philosophical heresies that surrounded the subject. He was, after all, a churchman as well as a chemist. That very year he had been appointed to the rectorship of the parish of North Runcton, near King's Lynn, an appointment that he would hold in conjunction with his

Cambridge professorship. Theologically orthodox, he certainly had no time for wild galvanic heresies.[21]

Thomas Verney Okes, the surgeon who took over to carry out the anatomical dissection of Weems' body once Cumming had finished with the battery, was something of an opportunist too. He had made quite a name for himself a couple of decades previously in the sensational case of Elizabeth Woodstock. Woodstock, a farmer's wife from Impington, just north of Cambridge, had leaped to national notoriety after surviving for eight days after falling from her horse and being buried in a snowdrift. Okes, the doctor who attended her once she was eventually discovered by a passing shepherd, published a pamphlet describing her ordeal and his treatment of her with mutton broth, wine and opium.[22] Elizabeth did not survive his ministrations for long, however, dying later that same year. The rumour was that strong drink had been the cause of her accident as well as her premature passing. With all that behind him, Okes by the 1810s had managed to acquire a reputation for 'steady judgement and practical skill'.[23] By the time of his encounter with Tom Weems, he was a prominent Cambridge medical man, married with an enormous family and an extensive and lucrative practice built around his house in Trinity Street. Until two years previously he had been a surgeon at the town's Addenbrooke's Hospital, before passing that plum position on to his son.[24] Weems probably looked like a prime chance to jump on the bandwagon of publicity yet again.

During the course of this very public series of gruesome experiments on that August afternoon in Cambridge, Cumming, Okes and their assistants had made full use of Smithson Tennant's large galvanic battery. In front of the capacity crowd that had gathered, crammed into the Chemical Lecture Room, they had worked their way systematically through Weems' body, trying to reproduce the motions associated with life. They had started by trying to make him breathe again and after that, moved on to try and stimulate facial expressions before finishing by attempting to make his arms and legs move around. Why were they doing this? What reasons did they have to suppose that a dose of electricity could bring a dead body back to at least the superficial appearance of animation? As we shall see soon enough, they were scarcely the first to try this. To make sense of their efforts and get a proper sense of just what had happened to Tom Weems' body, we need to pan back a little and see how this curious Cantabrigian experiment fitted into the broader scheme of things. We will need to leave Weems on the slab for a few chapters whilst we delve into the history and the politics of galvanism during the first few decades of the nineteenth century in Britain. When we eventually come back to Weems' body in the final chapter of this part of the book, we will see that there was nothing random about this episode. The experimenters on that August afternoon had some very specific goals in mind.

By the second decade of the nineteenth century, galvanism had already acquired a dangerous history and a dodgy reputation. It was at once the plaything of fashionable dilettantes, the hope of radical firebrands and the *bête noire* of conservative ideologues anxious to stamp out anything that seemed to smack of atheism, materialism and all such French connections. Playing with galvanism really did mean playing with fire. Barely a couple of decades previously the *Anti-Jacobin* had been busily and hilariously lampooning galvanists, along with chemists and other ne'er-do-wells, as dupes of the revolutionaries.[25] As the war with France was still a very recent memory, that kind of jibe remained fresh and painful. For Tories in particular, messing with galvanic piles was as bad as dancing with the devil. To rub salt in the wounds, Byron's *Don Juan*, published less than a year earlier, had also poked fun at the way 'galvanism has set some corpses grinning'. For the Cambridge experimenters, however, Weems' body offered the chance of making a name for themselves; it could also be turned into the perfect blunt instrument with which to beat radicalism over the head. What they really wanted, perversely, was for their experiment to fail – and that is exactly why it provides us with such a useful window through which to survey the early nineteenth-century galvanic scene.

TWO

GALVANISING BRITAIN

In successive editions of the *Opticks* – the second of his two famous books of natural philosophy, first published in 1704 – Isaac Newton introduced lists of 'queries' in which he asked questions about the nature of the universe. In one of these queries (query twenty-four, in fact) Newton speculated about the relationship between mind and matter. 'Is not Animal Motion perform'd by the Vibrations of this Medium,' he asked, 'excited in the Brain by the power of the Will, and propagated from thence through the solid, pellucid and uniform Capillamenta of the Nerves into the Muscles, for contracting and dilating them?'[1] When Newton asked a question like this, it was usually as a roundabout way of saying what he really thought. And for many of his readers throughout the eighteenth century, what Newton thought had all the certainty of gospel truth. If the man who had made sense of gravity thought something, then that was the way things were. The medium Newton had in mind was the ether – a subtle fluid that filled all of space. By the middle of the century, some natural philosophers speculated that electricity might also be an aspect of this universal medium (as Newton had similarly hinted). There was a link there to be made, licensed by no less an authority than Newton himself, between electricity and the operations of life. From that kind of perspective, it looked as if electricity might turn out to be a promising tool for addressing the abiding Enlightenment preoccupation with the ways minds, nerves and bodies worked; how thoughts got translated into actions; and what bodily mechanisms could tell us about the state of society.[2]

Seventeenth- and eighteenth-century natural philosophers produced electricity by rubbing a variety of substances, or 'electrics', such as amber and glass. When they were rubbed in the proper way with a piece of cloth, usually leather or something similar, these electrics were known to develop the power to attract or repel light objects like feathers or scraps of paper.[3] By the beginning of the eighteenth century, some of Newton's philosophical followers, such as Francis Hauksbee and Jean Théophile Desaguliers – the son of an émigré French Huguenot – had developed electrical machines that could produce this mysterious power in larger quantities. These machines typically consisted of a glass globe or cylinder, mounted on a wooden frame and arranged so that it could be rotated by turning a handle. When a cloth or even the operator's hands were held against the glass as it rotated, elec-tricity was produced. In 1729 the English natural philosopher Stephen Gray made the startling discovery that it was possible to make electricity's presence felt some distance away from the original source. He identified a class of substances, including metals (which eventually came to be called conductors) through which electric-ity could be communicated away from its source. One of these substances was the human body. One of Gray's favourite ways of demonstrating his discovery was to suspend a small child in the air with one of their feet in communication with an electrical machine. When the machine was operated, the child was electrified and could attract or repel small pieces of paper with their hands.

For followers of Newton, electrical experiments like these had a very serious purpose. They were the building blocks not just of a new natural philosophy, but also of a new theology. As Newton would have it, the phenomena of heat, light, electricity and magnetism were expressions of the active powers of God in nature. Newton's God was not some absentee landlord who had switched on the lights and then left the building. He was immanent in the universe – always there and always active. The powers of nature were not properties of matter, far from it; matter without those powers was dull and inert. These powers were added on to matter by God. They were His way of keeping the universe going. Newton had, in some ways, seen himself as being a sort of priest of nature, and many of his disciples also saw themselves in the same way, charged with the task of making God's powers visible. God (and Newton) had given them the task of producing 'ocular demonstrations' of 'that Fam'd Power in Matter, concerning which so much has been said, and so many noble and useful Discoveries have been made; *Attraction*, I mean, the Grand Principle which holds the whole Corporeal World together'.[4] Even John Wesley was impressed and left one such electrical machine performance wondering 'how a thin Glass Bubble, about an Inch Diameter, being half filled with Water, partly gilt on the outside, when electrified gives as strong a Shock as a Man can well bear?'[5] In fact, Wesley was so impressed that he went on to become a major advocate of the medical use of electricity.[6]

Demonstrations like these of the mysterious powers of electricity were very popular in fashionable eighteenth-century circles, though possibly for reasons sometimes less refined than those of the founder of English methodism and the more strait-laced Newtonians. There was a vogue for popular scientific lecturing in coffee houses and fashionable salons, and electricity was perfectly suited for the kinds of spectacular demonstrations that accompanied these lectures. Soon there was fierce competition to produce ever more eye-catching shows that could draw in crowds of 'ladies and the people of quality, who never regard natural philoso-phy but when it works miracles'.[7] Better machines and new instruments like the Leyden jar – which could be used to store electricity and release it in large quanti-ties – were the direct outcomes of this race for striking new demonstration devices. Joseph Priestley could not help but wonder:

[what] would the ancient philosophers, what would Newton himself have said, to see the present race of electricians imitating in miniature all the known effects of that tremendous power, nay, disarming the thunder of its power of doing mischief, and, without any apprehension of danger to themselves, drawing lightning from the clouds into an private room and amusing themselves at their leisure by performing with it all the experiments that are exhibited by electrical machines."[8]

Electricity could make Enlightenment ideas visible, but that, by the middle of the century, was double-edged.

Electrifying the human body often took centre stage in these displays of electricity's powers. Stephen Gray's experiments with the poor charity boys of Christ's Hospital suspended in mid-air are a case in point. They were living testimony that the electrical fire could be channelled through the human body. In another famous experiment, the German performer Georg Matthias Bose announced a spectacle he called beatification, in which a glowing halo could be made to appear above the head of a selected member of his audience. The secret of how to produce effects like beatification were jealously guarded and a cause of much contention, whilst other displays of electricity's powers were a little easier to reproduce. The Venus Electrificata was one particularly popular *pièce de résistance*. In this experiment, one of the more attractive female members of the audience would be invited to sit on an insulated stool and asked to hold a chain connected to an electrical machine. As the machine – and therefore the young woman – was charged, nothing appeared to happen. The gentlemen present would then be challenged to kiss the young lady. When one of them tried to take up the challenge, sparks would quite literally fly from one pair of lips to the other.[9]

One of the mid-eighteenth century's more innovative electrical performers was the Abbé Nollet. He was born Jean Antoine Nollet in 1700 to a peasant family and educated for the Church. By 1728 he had become a member of the *Société des Arts*, devoted to promoting new inventions. Thanks to his contacts there, he soon succeeded in getting his foot firmly in the door of France's philosophical elite, and by the 1750s he was one of the country's most fashionable scientific lecturers. Nollet had been inspired by Bose's German experiments. As he wrote excitedly to a friend in April 1745, he had heard Bose could electrify a man to such a degree that his 'hair … became luminous, which he jokingly calls *beatifying electricity*; that sparks from his fingers killed flies; that drops of his blood looked like drops of fire in the dark'.[10] He promptly set out to repeat the German experiments and before long was outdoing Bose with the scale of his demonstrations. Famously, Nollet entertained his king with a demonstration of electrical power in which a line of 180 royal guardsmen jumped simultaneously into the air when they received the shock from a battery of Leyden jars. On another occasion, he performed the same trick on a line of 200 Carthusian monks. It was a performance finely calculated to appeal to the absolutist monarch.

These kinds of extravagant demonstrations by Gray in England, Bose across the Rhine and Nollet in Paris, made it clear to the discerning eighteenth-century public that there was a particularly close connection between electricity and the body.[11] The issue seemed established beyond doubt when the Bolognese physician Luigi Galvani announced to the world in 1791 that he could produce electricity directly from animal tissue. Galvani had been experimenting on frogs' legs. According to legend, Galvani's wife Lucia (herself the daughter of Domenico Gusmano Galeazzi, Galvani's former teacher, patron and Professor of Physics at the University of Bologna) had been preparing the frogs' legs for her husband's dinner whilst a thunderstorm raged outside. Galvani happened to notice that the frogs' legs twitched convulsively with each lightning strike. Intrigued, he investigated further. He found that when a frog's leg's nerve and muscle were connected through a metallic circuit, the leg twitched. Galvani took this to be evidence of the existence of a distinct form of electricity, which he dubbed 'animal electricity'; similar but not identical to the common electricity derived from electrical machines and Leyden jars. Galvani argued that the brain was the source of this animal electricity, which was then conducted through the nerves to the rest of the body and stored in the muscles in the same way that ordinary electricity was stored in a Leyden jar.[12]

Galvani's claims promptly came under attack from a fellow Italian, Alessandro Volta, Professor of Physics at the University of Pavia. Volta had little time for ignorant medical men (as he took Galvani to be) dabbling in electrical experiments. His first assumption when he heard of Galvani's claims was that they were nonsense. After repeating the experiments, however, he found that the phenomenon Galvani described was real enough. Volta wanted nothing to do with any mysterious 'animal electricity' and set out to debunk Galvani's explanation. He claimed that the source of the electricity lay not in the frogs' legs, but in the metal strips that were used to complete the circuit. Galvani had noted that two different kinds of metal were needed to make the experiment work and Volta concluded that it was the contact of these two metals that was the real source of electricity. All the frogs' legs did was act as a conductor. The argument between the two savants raged back and forth for most of the following decade. After Galvani's death in 1798 others took up the cudgels on his behalf. Volta seemed to have the last word in 1800 when he announced to the world the invention of a new piece of apparatus that seemed to demonstrate his theory beyond question: the voltaic pile was made of copper and zinc discs in contact, each pair separated by cardboard soaked in acid, and it produced a steady stream of electricity.[13]

Volta seemed vindicated because his pile produced electricity without the presence of animal tissue. To rebut him, Galvani's followers needed to produce electricity from flesh without the presence of metals. They proceeded to do this with some style. In August 1802 a trio of Turinese experimenters, Giulio, Rossi and Vassalli-Eandi, reported on the results of some experiments carried out on the corpses of recently executed criminals in the city. To confound their enemy they needed to find ways of demonstrating, conclusively, the continuing sensitivity of human flesh to electricity, and also show that it was itself a source of electricity. They reported how, by 'arming the spinal marrow by means of a cylinder of lead introduced into the canal of the cervical vertebrae, and then conveying one extremity of a silver arc over the surface of the heart, and the other to the arming of the spinal marrow', they managed to successfully produce some 'very visible, and very strong contractions' of the corpse's heart. They made a point of emphasising, in case their readers missed the crucial point, that the 'experiments, as seen, were made without any intervention of the pile, and without any armature applied to the heart'.[14] To all appearances, the electricity producing these contractions had somehow emanated from the corpse itself.

Galvani's nephew, Giovanni Aldini, took the challenge even further, embarking on a grand tour of the war-torn continent to defend his uncle's philosophical reputation against Volta's calumnies. He followed Volta to Paris in August 1802 to show off his experiments before the First Class of the National Institute (the Republican successor to the Royal Academy of Sciences). Using the heads of decapitated oxen he demonstrated to the sceptical savants – who had, after all, just welcomed Volta into their midst and awarded him a gold medal at Napoleon's behest – that electricity could indeed be produced from animal bodies.[15] One spectator left the demonstration 'charmed and transported with admiration at the simplicity of the means which nature employs in its phenomena that seem to us the most complex'. It seemed to this willing convert that galvanism accounted for 'a great number of the phenomena of the animal and vegetable kingdoms'. He agreed with Aldini that nerves and muscles were organised as galvanic piles, and that they were arranged in animal bodies in such a way that 'they discharge, in regard to each other, the same functions as the different metals the contact of which excites a permanent current of the electric fluid; which is the most valuable discovery for which we are indebted to the pile of Volta'.[16]

Later that year Aldini repeated his experiments at Oxford and before the Royal Society in London, taking advantage of the short-lived Peace of Amiens to cross the Channel. In London Aldini performed his experiments at the Great Windmill Street Anatomical Theatre, where the celebrated surgeon John Hunter had once plied his trade. They were later repeated in front of the gathered surgeons and medical students of Guy's and St Thomas' Hospital, with Astley Cooper, one of the metropolis' most flamboyant and eminent bone-cutters, presiding. Aldini's experiments with frogs, decapitated dogs and rabbits received tumultuous applause and the triumphant performer found himself presented with a gold medal commemorating his achievements.[17] Some, at least, of the Royal Society's fellows were suitably impressed; one of them enthused:

Here then we have the most decided substitution of the organised animal system in the place of the metallic pile: it is an animal pile; and the direct production of the galvanic fluid, or electricity, by the direct or independent energy of life in animals, can no longer be doubted.

There was no doubt in their minds that 'Galvanism is by these facts shewn to be animal electricity; not merely passive, but most probably performing the most important functions in the animal economy'.[18]

Whilst in London, Aldini was also performing to audiences already pre-disposed in his favour. Galvani's experiments had captured the imagination of natural philosophers already interested in investigating the connections between electricity and the nervous fluid. Ironically enough, Volta's invention of the voltaic pile and his challenge to see off animal electricity had only served to keep Galvani's ideas alive and kicking. In Britain, natural philosophers rushed to try Galvani's and Volta's experiments for themselves. When Humphry Davy, the new doyen of metropolitan natural philosophy, arrived at the Royal Institution from his pupilage as Thomas Beddoes' chemical assistant at the Bristol Pneumatic Institute, he took up Volta's new instrument with gusto.[19] Other natural philosophers, doctors and instrument-makers, such as John Cuthbertson, William Nicholson and Joseph Carpue (who was to assist Aldini with some of his public experiments in London), were industriously tinkering with their apparatus as well, improving Volta's design and looking for new experiments to impress the public and their fellow philosophers. Natural philosophical journals, gentlemen's magazines and newspapers alike were awash with news of the latest developments in galvanism. Nicholson, in his *Journal of Natural Philosophy* and Alexander Tilloch in his *Philosophical Magazine* rushed to print the latest galvanic news from the continent.

The grand climax of Aldini's English tour came in January 1803 when he got the opportunity to experiment again on a human body (he had already had one opportunity at home in Bologna a year or so earlier). On 17 January 1803, George Forster was hanged at Newgate for murder. Following his execution his body was handed over to the College of Surgeons for dissection and Aldini got his chance to experiment. Before an audience presided over by Thomas Keate, President of the College of Surgeons, Aldini put Forster's remains through their paces using a pile of 120 plates of zinc and copper. The result was a grand display of contractions and convulsions:

On the first application of the process to the face, the jaw of the deceased criminal began to quiver, the adjoining muscles were horribly contorted, and one eye was actually opened. In the subsequent part of the process, the right hand was raised and clenched, and the legs and thighs were set in motion. It appeared to the uninformed part of the by-standers as if the wretched man was on the eve of being restored to life.

In reality, this was a rather unlikely eventuality, since as the commentator on this occasion dryly remarked, 'several of his friends, who were under the scaffold, had violently pulled his legs, in order to bring a more speedy termination to his sufferings.'[20]

Forster was described at the time as 'a decent looking young man' and he had apparently 'died very easy'[21] (such things were presumably considered to be relative). After hanging, his corpse had also been kept in cold conditions for a period; he was therefore, one imagines, in reasonably good condition for Aldini's attentions by the time he arrived on the dissecting table. The battery for the performance was relatively large, 120 plates of copper and zinc was a significant experimental resource for the period and not commonly available. It was more than twice the size of the battery used a few months previously in Turin by Vassali-Eandi, Rossi and Giulio in their electrical dissections of a couple of decapitated human bodies. It also had twice as many plates as the 'most powerful Galvanic apparatus' constructed for William Haseldine Pepys at about the same time.[22] There is no record of its exact design or where Aldini acquired it – he is very unlikely to have brought it with him all the way from Bologna – but it may be significant that the instrument-maker John Cuthbertson, celebrated for his powerful electrical machines and an avid improver of the voltaic pile, was one of Aldini's fellow experimenters on this occasion.

Aldini's performances were a minor sensation and were patronised by London's fashionables, distracting even the Prince of Wales from his usual frolics. They attracted the interest of the Royal Humane Society and electricity was touted as a possible way of saving the victims of drowning. Not only doctors and natural philosophers had flocked to see his experiments; when Aldini performed at the Great Windmill Street Anatomical Theatre his audience included General Andreossi, the French ambassador, along with his entourage; Argyropoli, the *chargé d'affaires* of the Ottoman Porte; and the ubiquitous antiquarian, Sir William Hamilton. A subsequent performance was attended by no fewer than four peers of the realm, including the rising political star, Lord Castlereagh. This, along with the portly Prince George's interest, shows how galvanism played in polite culture. The dissolute George certainly had not the slightest interest in the politics of galvanism (though the canny Lord Castlereagh might). He was there simply to be amused. For onlookers like this, performances such as Aldini's were simply one more in the range of entertainments the capital city offered. They went to see Aldini's electrical dissections in much the same spirit as they went to the theatre, were pleasurably titillated at the phantasmagoria or wowed by the latest panorama.[23] Though Aldini, writing about the experiment well over a decade later, described it as 'a prostitution of galvanism, if it were only employed, to cause sudden gestures, and to convulse the remains of human bodies, as a mechanic deceives the common people by moving an automaton by the aid of springs and other contrivances',[24] it seems hard to resist the conclusion that this was partly what he was engaged in doing with what was left of George Forster. He did, after all, need to draw attention to his performance.

In this respect, if in no other, there was nothing even remotely radical about Aldini's experiments. On the contrary, they were conducted with the full blessing of the English state. When Aldini performed his galvanic dissections on the body of George Forster, he was not only doing so with the full approval of the English government, he was quite explicitly performing a role that was designed to underline and reinforce the judicial power of the state. His experiments on poor Forster were the climactic last act in a performance that had commenced with the conviction, proceeded with the brutal public execution and concluded with the eventual dismemberment of the criminal's body once Aldini was done with it.[25] The experiments were, moreover, carried out in defence of the theories of Luigi Galvani, who had been sacked from his professorial position in Bologna for refusing to swear the oath of allegiance to Napoleon following his conquest of Italy, and was the staunch enemy of the man who had so recently been lionised in Paris by the First Consul for his invention of the voltaic pile.[26] There was, nevertheless, a dangerous undercurrent to these performances which investigated the boundaries between the living and non-living. Many Englishmen – Edmund Burke amongst them – listed galvanism amongst the rogue's gallery of malign forces responsible for the French Revolution.

Experiments and ideas that appeared to link electricity to the body were a source of endless inspiration to radical political agitators demanding the rights of man. Any kind of philosophical theory that smacked of materialism – the notion that there was no more to life than the movement of organised matter – was anathema to the politically orthodox. The radical natural philosopher Joseph Priestley – whose house and laboratory in Birmingham had been burnt to the ground by a Church and King mob, baying for his blood only a few years previously in 1791 – had argued that 'the English hierarchy (if there be anything unsound in its constitution) has equal reason to tremble, even at an air-pump or an electrical machine'.[27] He meant that electricity exposed the iniquities of a corrupt state by revealing the true order of things. Others, like John Thelwall, concurred. Thelwall, radical rabble-rouser, poet and founder of the deeply subversive London Corresponding Society, turned to electricity to argue for the material basis of life and the rights of man as an inevitable consequence of natural law. In response, magazines like the *Anti-Jacobin* and caricaturists such as James Gillray held Priestley, Thelwall and their ilk to ridicule, lampooning them furiously and savaging their enthusiasm for galvanism along with the rest of their revolutionary and unpatriotic creed.

The radicals drew much of their materialist inspiration from the infamous Frenchman, Julien Offrey de la Mettrie and his blasphemous outpourings in *l'Homme Machine* in 1748. Man really was just a machine, de la Mettrie thundered, without even the benefit of the soul that the great French philosopher Descartes had allocated him. Inspired in his turn by the creations of engineer Jacques Vaucanson, famous for his mechanical automata, de la Mettrie held that humans were no more than clockwork toys that were somehow able to rewind themselves. He insisted:

> The soul is therefore but an empty word ... of which no one has any idea, and which an enlightened man should use only to signify the part in us that thinks. Given the least principle of motion, animated bodies will have all that is necessary

for moving, feeling, thinking, repenting, or in a word for conducting themselves in the physical realm, and in the moral realm which depends upon it.[28]

Elsewhere he speculated that the 'human body was made up of nerves, stretched to a certain point, filled with spirits which circulate with the prodigious activity of fire', and that 'the very fire that comes out of our bodies in the form of electricity goes to make up these spirits'.[29] No soul, of course, meant no God – and no God meant no pre-ordained social order either.

If materialism was a problem, German philosophers were jumping to the opposite – and equally politically dangerous – extreme. *Naturphilosophen*, such as Johann Wilhelm Ritter, were loudly proclaiming how galvanism demonstrated that the entire cosmos was a single living being – a proposition as destructive to conservative religious and political orthodoxy as the most arrant French materialism. *Naturphilosophie* had developed in the German states as an explicit antidote to what its promoters and supporters regarded as the excesses of Newtonian philosophy. They abhorred what they saw as Newton's mechanistic, clockwork and soulless universe, (though Newton himself would have regarded his view of the universe as neither mechanistic nor soulless) and embraced instead a view of the cosmos as an animate entity. According to Friedrich Schlegel, 'if you want to penetrate into the very core of physics, have yourself initiated into the mysteries of poetry'.[30] It was a sentiment typical of this new German view of natural philosophy as an individual, intimate endeavour. Its proponents jumped enthusiastically on the galvanic bandwagon, recognising that Galvani's experiments were a powerful argument in favour of their belief in an all-pervading animating fluid, flooding and vivifying the entire universe. *Naturphilosophen* like Ritter – or the grand old German poet Goethe – were certainly not radicals. They rejected utterly the extreme materialist views that were taken to underpin late eighteenth-century radical political culture.

The rejection of the moderate Anglican God of orthodox Newtonian theology looked dubious indeed to many British natural philosophers. 'Where is the sun, where is the atom that would not be part of, that would not belong to this organic universe, not living in any time, containing any time?' speculated Ritter. 'Where then is the difference between the parts of an animal, of a plant, of a metal, and of a stone? – Are they not all members of the *cosmic-animal*, of *Nature*?'[31] This was straying on to dangerous theological (and therefore political) territory as far as many of his contemporaries in England were concerned, particularly as it turned out that the pantheistic tendencies of these heterodox philosophers' views of nature were all too easily co-opted by British radicals as well. The *Naturphilosophen* certainly had their British sympathisers. The brash young poet Samuel Taylor Coleridge was one of them. In 1798 he set out for Göttingen with the Wordsworths in tow, to learn the language and to imbibe the true German philosophy by sitting at the feet of the original masters. Coleridge would infect his friend, the rising scientific star Humphry Davy, with this dangerous 'German disease' as well – at least for a little while.

THREE

GALVANIC FASHIONS

At the beginning of the nineteenth century it seemed indisputable to many natural philosophers that there was a firm connection between electricity and animal and human bodies. As far as they and the savants were concerned, Galvani and his combative nephew Aldini had successfully established the existence of animal electricity. Experiments such as Aldini's flamboyant demonstrations with body parts, dead animals and executed criminals appeared incontrovertible. Electricity really could be produced from living (or recently living) tissue and Volta had been wrong to suppose that it was produced by the contact of metals alone. There really did seem to be a 'galvanic fluid peculiar to the animal machine, independently of the influence of metals, or of any other foreign cause'.[1] One French natural philosopher had even been successful in constructing a voltaic pile made entirely from what sounds rather like the sweepings from the anatomical theatre floor. Lagrave had found that 'by placing upon each other successive layers of muscular fibre and of brain, separated by a porous body, soaked in salt water, a pile was formed which produced the usual effects of the galvanic apparatus'.[2] This was powerful evidence – as far as Aldini's supporters were concerned at any rate – not only that electricity could be produced from animal tissue, but that there was some sort of intimate connection between galvanism and the processes of life.

It remained open to dispute just what the role and nature of such animal electricity was, however. There was also disagreement over what the relationship was between galvanism and the common electricity produced by electrical machines. After all, in many respects the effects produced by galvanism and common electricity were very different. Aldini's experiments may have seemed convincing to his many followers, but they had certainly not convinced everybody. There were still those who believed, with Volta, that animal electricity was a chimera and its practitioners either deluded or deceivers. Even some of Aldini's supporters, like the surgeon and galvanic lecturer Charles Wilkinson, doubted some of the experimental evidence he had gathered. Wilkinson pointed out that he and other sympathisers had failed to reproduce some of the results that Aldini described. Even when they accepted the results, not all were sure that they added up to as convincing a body of evidence as Aldini wished them to believe.[3] Where there was agreement about the existence of animal electricity, there was still considerable disagreement over the precise nature of the galvanic fluid's relationship to the nervous power and the processes of life. Natural philosophers vied with each other in trying to produce convincing and plausible accounts of just how galvanism fitted into the natural economy.

London, in particular, sustained a thriving and lively culture of electrical experimentation and instrument-making in the opening decades of the nineteenth century. Throughout the eighteenth century, electricity had been widely regarded as a way of making the Enlightenment visible. Impressive displays of sparks and electric fires were guaranteed to bring in the crowds and provide tangible proof of the performer's intimate mastery over the powers of nature.[4] Galvanism became an integral part of this culture of display, providing a ready source of new and spectacular demonstrations. Rival philosophers and instrument-makers competed avidly with each other to make the biggest and the best. By 1803 William Haseldine Pepys was already boasting his possession of 'the most powerful Galvanic apparatus that has, we believe, been yet produced'.[5] It consisted of a huge array of sixty pairs of zinc and copper plates, arranged in two large troughs filled with acid. Instrument-makers such as William Cruickshank (on whose design Pepys' battery was based) and John Cuthbertson were busy finding ways of making galvanic arrangements that were more powerful and produced increasingly spectacular and, quite literally, shocking displays. The human body itself – that of the performer and those of his audience – was an important component in this galvanic economy of display. It could be treated as a galvanic instrument and played with to find the best ways to show off the galvanic fluid.

There was no question, however, as to whom the consummate showman was dominating London's galvanic scene. Humphry Davy at the Royal Institution ruled the roost. The Royal Institution, where Davy reigned as Professor of Chemistry, had been established as a typical Enlightenment endeavour, designed to use science for the betterment of humankind. The man behind the Royal Institution, Benjamin Thompson, was a loyalist refugee from the lost American colonies who had been contriving to make a name for himself (as well as acquiring the title of Count Rumford from the Elector of Bavaria along the way) with a variety of visionary schemes of varying practicality. With patronage from Sir Joseph Banks, the Royal Society's then despotic president, he acquired sufficient financial backing to launch the Royal Institution with its patrician premises on Albemarle Street in 1799 as an establishment dedicated to improvement.[6] It had been part of the Royal Institution's original conception that there should be room for the lower orders, hungry for instruction in chemistry and mechanics, in its lecture theatre, as well as for the toffs. However, that benevolent intention did not survive the trauma of the revolutionary wars and the political lurch rightwards that accompanied them, the idea abandoned as having a 'political tendency'.[7]

By the time Davy arrived there, the place was already well on the way to becoming what the essayist Thomas Carlyle acerbically described as a 'kind of sublime mechanics' institute for the upper classes'.[8] Davy had a great deal to do with completing this transformation with his spectacular galvanic performances; designed to titillate and amuse the jaded fashionables in his audience. He had moved quickly to cast off past associations with dangerous radicals like his old patron Thomas Beddoes at Bristol, and enthusiastically embraced the Institution's aristocratic science. Beddoes had been a staunch advocate of radical science and sensational self-experimentation, as indeed Davy had been during his Bristol days.[9] He and his patron had embraced galvanism and its possibilities with gusto. In one of his first effusions as Beddoes' new protégé, Davy took for granted the identity of the electric fluid and the nervous influence, arguing that electricity itself was in fact 'condensed light', providing 'another cogent reason for supposing that the nervous spirit is light in an ethereal gaseous form'. Life, in that case, was 'a perpetual series of peculiar corpuscular changes', and 'perceptions, ideas, pleasures, and pains, are the effects of these changes'.[10]

With radical Bristol behind him and the aristocratic Royal Institution beckoning however, Davy ditched this radical past with some alacrity. In notes added later to his 1799 experimental notebooks, Davy drew a clear line between his present and former selves. He had commenced his 'pursuit of chemistry' with 'speculations and theories: more mature reflection convinced me of my errors, of the limitations of our powers, the danger of false generalisations, and of the difficulty of forming true ones'.[11] By 1801 in 'Outlines of a View of Galvanism', one of his first publications in the *Journal of the Royal Institution*, Davy would only say that 'some phenomena similar to galvanic phenomena, may be connected with muscular action, and other processes of life'. In 1803 Davy contributed an anonymous review of Aldini's *Account of the Late Improvements in Galvanism* to the recently established and decidedly Whig-leaning *Edinburgh Review*. Writing that the dross of early and unfounded galvanic speculation needed to be swept aside, he asserted: 'every new light thrown upon natural knowledge, at first dazzles and confuses … it is only by degrees that the just appearances of the objects of discovery are perceived, and their relations ascertained.' In Davy's view, Aldini was clearly still dazzled: 'His reasonings on this subject appear to us to be very inconclusive indeed.' As for Aldini's experiments on Forster's corpse, they were simply 'rather disgusting than instructive'.[12]

Rather than continuing to play with bodies, Davy put the Royal Institution's huge galvanic batteries to good use, in the laboratory as well as the lecture theatre. He churned out new galvanic discoveries and new chemical elements that could be used to subvert the recently popular French system of chemistry, established by the late Antoine-Laurent Lavoisier.[13] With revolutionary war fever in the air, any stick that could be used to beat the pompous French systematisers on the other side of the Channel – even one like Lavoisier who had fallen foul of the Revolution's guillotine – was to be welcomed. Davy made such a name for himself as Europe's premier chemist and galvanist that in 1813 he even received special dispensation from Napoleon himself (anxious to appear above mere national feeling) to travel to Paris and receive the emperor's accolade for services to science. In tow with him he took his young apprentice, Michael Faraday. Davy used the opportunity to rub the French savants' noses in the superiority of English chemistry, dismaying them with his galvanic batteries' ability to probe into the mysteries of nature. It was this performance that went down well with his fellow philosophers back in London.

Despite Davy's dominance of the galvanic scene, the combative atmosphere bred rival theories of galvanism which flourished and multiplied. Agreement was widespread that galvanism appeared to have some particular close relationship with the functioning of the body. As Charles Wilkinson remarked, 'all fluids yet known, except air and oil, contain more or less electricity, and will freely allow its ingress, as well as egress. As the human body is principally constituted of fluids, it is replete with electricity, and sensible to the least disturbance.'[14] But recognising the existence of electricity in human bodies was a long way from agreeing what it was doing there. There was significant disagreement as well over just how the nervous system operated; how nerves transmitted movement and sensation from brain to body and back again. The Enlightenment philosopher David Hartley had suggested that nerves were like tightly stretched ropes and that messages passed back and forth through vibration. Another popular theory was that this took place by means of a specific nervous fluid, and many argued that this nervous fluid and galvanism might well be the same thing.[15] After all, if electricity was incontrovertibly being produced in the body, then it clearly performed some function as well. But even amongst those who agreed on the identity of galvanism and the nervous fluid – or at least that there was a strong analogy between them – there remained large scope for disagreement.

Wilkinson, for one, argued for a special relationship between galvanism and the nervous fluid. For him, galvanism was 'an energising principle, which forms the line of distinction between matter and spirit, constituting in the great chain of the creation, the intervening link between corporeal substance and the essence of vitality'.[16] Medical galvanist Matthew Yatman concurred, agreeing that the 'animal vital principle, formerly called "The nervous fluid" is the connecting medium between mind and body'; he was in no doubt that this was electrical in nature.[17] But even Wilkinson had to admit that claims like this really did very little towards establishing just how galvanism operated on the body. As he put it:

On the supposition that galvanism is the intermediate principle between matter and spirit, I cannot, I must confess, conceive the mode in which the agency is effected. To comprehend the essence of our own animation, requires the powers of a principle superior to that we possess. Infinite as I regard the difference between common matter and our vital principle, still we may suppose another infinitude, between our spring of life and that source that comprehends all.[18]

Arguments for the identity of electricity and vitality did nothing to establish just what those mysterious principles really were.

Wilkinson's was nonetheless just the kind of talk to get the poets interested. The little coterie of poets and writers surrounding Samuel Taylor Coleridge and Robert Southey were fascinated by galvanism. Humphry Davy had himself been an important member of this group during his radically-tinged years as Thomas Beddoes' chemical assistant at the Pneumatic Institute near Bristol. Coleridge dabbled (or tried to dabble) in chemistry whilst Davy, with a little more success, turned his hand to poetry. Beddoes had established the Pneumatic Institute as a base to investigate the therapeutic uses of different gases. It had been Davy's task to analyse the potential of these gases as therapeutic agents and oversee their use on the Institute's clients. Coleridge and company, who spent considerable time in the neighbour hood of Bristol and the nearby Quantock Hills during the 1790s, embroiled in a number of doomed efforts to set up a democratic and self-sustaining community of radical writers and were fascinated by Beddoes' and Davy's project. They regarded it as a grand experiment investigating the relationship between mind and matter, looking at how corporeal substances like gases affected the workings of spirit. Indeed a dose of laughing gas was a common treatment at the Pneumatic Institute. It was for precisely these sorts of reasons that they were intrigued by galvanism.[19]

As we have seen, Coleridge was fascinated by the new *Naturphilosophie* emerging in the German states, too. He had travelled extensively there in 1798 and 1799, spending much of his time in the university town of Göttingen. He used his time there diligently, attending the lectures of eminent German scholars such as the naturalist Johann Friedrich Blumenbach. The University of Göttingen at the end of the eighteenth century was a major German centre of intellectual culture, and Coleridge had immersed himself with enthusiasm in the town's heady atmosphere. By the time he eventually returned to England he could boast: 'I have learnt the language, both high & low German … I have attended lectures of Physiology, Anatomy & Natural History with regularity, & I have endeavoured to understand these subjects.'[20] The emphasis on science was significant, and engaging with the latest German ideas in natural philosophy had been one of Coleridge's main motivations for his German tour. He was much taken with its emphasis on the transcendental unity of things and the emphasis on the imagination as a legitimate tool of philosophical and scientific inquiry. He admired the way that the Germans appeared to make no distinction – in his eyes at least – between the study of nature and the study of self.[21]

For the same reasons, Coleridge was taken with galvanism. Both he and Robert Southey were enthused by the latest discoveries and by their friend Humphry Davy's involvement. Southey speculated to Davy that Volta's experiments showed that 'as the galvanic fluid stimulates to motion, that it is the same as the nervous fluid; and your system will prove true at last'.[22] Galvanism had clearly been a topic of some discussion between them already. When Coleridge learned that his friend was

about to lecture on galvanism at the Royal Institution, he wrote to Davy that his
'motive muscles tingled and contracted at the news, as if you had bared them, and
were *zincifying* their life-mocking fibres'.[23] Beneath the playful ribbing, Coleridge's
remark to Davy was far more than mere frivolity. It showed that he knew something, at least, about the details of galvanic experimentation. And Coleridge really
did think that galvanism was a good way of trying to understand how poetic imagination and inspiration worked. In his view it all boiled down to the question of
how spirit interacted with matter. Galvanism appeared to him, as it did to doctors
like Charles Wilkinson, like a power that could mediate between body and soul.

By the beginning of the nineteenth century electricity was well-established as a
form of medicine. Given the widespread interest in the notion that there was some
intimate link between electricity and the vital forces of life, this was hardly surprising. At its simplest, the idea was that a good healthy dose of electricity could restore
the depleted vital fluids that had led to ill-health in the first place. Luigi Galvani's
discoveries and the consequent surge of interest in galvanism only added plausi-
bility to the view that electricity could cure disease. Medical entrepreneurs were
certainly keen to cash in on electricity, and lecturer and medical practitioner James
Graham is a good example of this. After studying medicine in Edinburgh, Graham
spent time in North American colonies where he became familiar with Benjamin
Franklin's spectacular electrical experiments. Appreciating their potential, and with
an eye to the main chance, he returned to England and set himself up as a purveyor
of electrical medicine in fashionable Bath, advertising himself as a specialist in nervous disorders. Moving to London in the late 1770s, he established his Temple of
Health on the Strand, where clients paid up to £50 a go for a night on his Celestial
Bed to cure impotence and infertility through electricity. The connection Graham
forged between electricity, sexual health and fertility was to last throughout the
nineteenth century.[24]

Edinburgh, where Graham had studied (but not taken a degree) was one of
eighteenth-century Europe's main centres of medical learning. It soon became a
centre for interest in galvanism and its medical applications too. Alexander Munro
secundus (the second in a dynasty of Alexander Munros who dominated the teaching of anatomy at Edinburgh in the late eighteenth and early nineteenth century)
took a keen interest in Galvani's discoveries and their possible medical uses. Others
at Edinburgh, such as Richard Fowler and even the rabidly anti-radical Professor
of Natural Philosophy, John Robison, were keen galvanic experimenters. By the
first decade of the nineteenth century galvanism was clearly also on the menu elsewhere; for medical students at London's schools of medicine, for example. Charles
Wilkinson, surgeon and lecturer in experimental philosophy at St Bartholomew's
Hospital, lectured on galvanism at the Soho Square Anatomical School. Joseph
Carpue gave lectures on galvanism and medical electricity to the students attending his lectures on medicine and anatomy at Dean Street. The surgeon John Birch
was an enthusiastic practitioner of medical electricity at St Thomas' Hospital until

his death in 1815 (though he confined himself to using the old-style frictional electrical machines rather than embracing the new technology of galvanism). In addition, interest in galvanism amongst medical students and doctors alike at Guy's Hospital seems clear from the enthusiastic reception that Aldini received when he performed there.

Aldini himself was also, of course, interested in the medical applications of galvanism. He was a strong advocate of galvanism as an agent for resuscitating the victims of drowning or asphyxia. This was one of the reasons the Royal Humane Society (established in 1774 with the aim of investigating ways of medically resuscitating the victims of drowning) were so interested in his experiments. Certainly one of the aims of his experiments on the body of Forster was to investigate the possibility of returning him to life. The surgeon and political radical Joseph Carpue, who assisted Aldini in his electrical dissection of Forster, recorded how they tried to get the hanged man's heart beating again:

> Mr. Cuthbertson and myself immediately, by the desire of Sig. Aldini, applied the
> conductors to the heart, Mr. C. to the right ventricle, and I to the left. I exclaimed,
> 'the heart acts'; one of the gentlemen present thought it did. Sig. Aldini, Mr. Keate,
> Dr. Pearson, Mr. Cuthbertson, and the other gentlemen were of opinion it did not.[25]

In fact, the other gentlemen thought that Carpue had given the heart an inadvertent poke with one of the conductors. Aldini's failure to get Forster going again did not discourage him from further experiments nor did it reduce his faith in the therapeutic virtues of galvanism. Medical men such as Wilkinson, however, were already warning about the dangers of the indiscriminate use of galvanism in medicine and insisting on the importance of proper medical training: 'The application of so important an agent to the animal economy should not be left to the discretion of one who only comprehends the management of a machine, no more than the treatment of serious disease should be confided to one whose knowledge is limited to the mere compounding of medicines.'[26]

A thriving market was already developing in electrical and galvanic cures that anybody could use. During the 1790s an American doctor, Elisha Perkins, enthused by Galvani's experiments, started tinkering on his own account, using his patients as experimental subjects. The result was Perkins' Tractors – sets of metal rods that worked by redirecting the galvanic influence around the body. The tractors were immensely popular in the new republic and soon arrived in the Old World too, being introduced to the British Isles by Perkins' son, Benjamin. Benjamin Perkins set up shop in Leicester Square, offering personal treatment and sets of tractors for sale for 5 guineas. A healthy copy-cat market developed following Perkins' failure to protect his investment with a patent. Matthew Yatman was one of many who advertised his own version of the tractors for sale and for treatment. As far as most medical practitioners and galvanic enthusiasts were concerned, the tractors were arrant quackery and the rationale for their operation entirely spurious. Galvanists denied that they could either produce electricity or redirect the galvanic fluid around the body as Perkins and his followers suggested. Their huge popularity testified that many others felt differently. It also testified to the degree to which galvanism during the first few decades of the nineteenth century was coming to be recognised as something of a universal cure-all.[27]

Perkins' tractors were not the only pieces of galvanic medicine for sale on the streets of London and elsewhere. There was soon a thriving trade in galvanic belts, rings and other impedimenta that would last throughout the nineteenth and into the twentieth centuries. Their popularity testifies the degree to which galvanism had permeated Georgian society. Electricity had long been a part of fashionable culture in the metropolis and in provincial watering holes such as Bath – hence James Graham's decision to establish his practice in the town. Electrical machines even adorned the drawing rooms of polite households, to be hauled out on occasion for party games such as the Venus Kiss, or to allow the paterfamilias to demonstrate his erudition and familiarity with the latest thing.[28] Galvanism's intimate relationship to the body and its workings was common knowledge in such circles. Just as familiar was the knowledge that galvanism had a darker side as well. It might be the plaything of fashionables, but it was a tool for revolutionaries too. Speculation about what electricity did to the body and dabbling in galvanic medicine by buying an electric belt or trying out a set of Perkins' tractors was, according to some critics at least, just the first step on the slippery slope that led to materialism and godlessness.

Aldini himself had already embarked upon sliding down this slippery slope, according to some of his enemies, and for Tory satirists, Aldini's experiments were symptoms of delusion. Christopher Caustick's satirical defence of Elisha Perkins and his tractors in *Terrible Tractoration* (in reality written by the American writer Thomas Green Fessenden, who was acting as Perkins' agent in London) suggested that Aldini and his like certainly had no claims to superior virtue or knowledge over his employer. If Perkins was a quack, then so were they.[29] Aldini and galvanism made excellent targets:

> For he ('tis told in publick papers)
> Can make dead people cut droll capers;
> And shuffling off death's iron trammels,
> To kick and hop like dancing camels.
>
> To raise a dead dog he was able,
> Though laid in quarters on a table,
> And led him yelping, round the town,
> With two legs up, and two legs down.[30]

Caustick's doggerel assembled a veritable rogue's gallery of atheists, charlatans and materialists and explained exactly why Aldini belonged to this disreputable pantheon. John Corry had similar fun at Aldini's expense, marvelling at those 'most wonderful *distortions*' that could be produced by galvanism, and gleefully relating how:

> in an experiment made on malefactor who was executed at Newgate, he imme-
> diately opened his mouth – doubtless, another application would have made him
> speak; but the operators, Aldini, Wilkinson, and Co. were so much affrighted that
> they threw down their instruments and took to their heels.[31]

All of this galvanic speculation was certainly familiar enough territory to the liter-ary world of Byron, Polidori and the Shelleys as they lounged about in Byron's rented villa on the shores of Lake Geneva, swapping ghost stories as a storm raged outside. The story of how the young Mary Shelley came to write *Frankenstein* is no doubt a familiar one. Mary had secretly married and eloped to the continent with her lover, the brash radical poet Percy Bysshe Shelley. She came from a radi-cal and deeply intellectual background herself. Her father, William Godwin, was a notorious radical philosopher and author. Her mother, Mary Wollestoncraft, was equally infamous for her authorship of the *Vindication of the Rights of Women*. Percy Bysshe Shelley had dabbled with galvanic experiments as a schoolboy and student, and both he and Mary were perfectly familiar with Coleridge's writings and gal-vanic speculations. Byron, in addition, was quite familiar with the latest galvanic speculations and crazes. He and the Shelleys knew Coleridge and admired his work, though like many of the new generation of intellectuals and poets, they were all to different degrees horrified by what they regarded as his betrayal of the radical prin-ciples that had underpinned his earlier writings. There was plenty for Mary to play with, therefore, as she offered her own contribution to the ghost story competition. A tale about the possibilities and dangers of artificial life made perfect sense for the period and the company, as well as providing rather a neat vehicle for probing some of the real uncertainties of Regency society.[32]

The details of Mary Shelley's tale reveal just how familiar she was with the cul-tural minutiae of Georgian galvanism. Her hero, Victor Frankenstein, gained the knowledge that made his creation of the monster possible through study at a German university, where the Professor of Chemistry taught him how modern philosophers could 'command the thunders of heaven, mimic the earthquake, and even mock the invisible world with its own shadows'.[33] The monster was cobbled together from body parts purloined from graveyards and charnel houses – evocative and troubling in a society where body-snatchers flourished to serve the burgeoning medical market for the dead. Frankenstein's researches had delved into the myster-ies of how life was produced from seemingly inanimate matter. But after finding the secret of how to 'infuse a spark of being into the lifeless thing' he had cre-ated, Frankenstein rejected the monster he had made, recoiling in horror from his masterpiece. Mary Shelley's tale may have been fanciful, but it was well grounded in its borrowings from galvanic theory and practice. There was nothing in the natural philosophical details of her story that would not have seemed familiar to her readers. Victor Frankenstein's revulsion at the sight of his monstrous creation also evoked real contemporary concerns about what galvanic experiments might have to say about materialism, politics and society.

Galvanism might therefore mean any of a number of things for the sort of people who read and enjoyed *Frankenstein*. It was politically heterodox and dangerous. It was part of fashionable and consumer culture. It was inextricably wound up with anticipations and concerns about life, sex and death. Aldini's experiments seemed to show — on some readings anyway — that bringing the dead back to life through electricity was a real possibility. The radical Joseph Carpue certainly thought so. One way of making sense of that would be to acknowledge with the radicals that there was nothing more to life than organisation. But other galvanists like Charles Wilkinson read the relationship between galvanism and life very differently. For them, galvanism's links with vitality made it immaterial rather than making vitality material. In the meantime, galvanism for the likes of Perkins and Yatman represented a fine opportunity to make a fast buck by selling electrical cures to the affluent middle classes. Or was it just another fashionable fad? When Byron knowingly quipped in *Don Juan* that 'galvanism set some corpses grinning', it was a throwaway aside that he could be confident would be understood by his readers. And he could be just as confident that it would be read in any number of different ways — irreverently, ironically or, at the other end of the scale, with a frown and a scandalised mutter.

FOUR

BODY & SOUL

England during the years of the Napoleonic Wars was a political cauldron seething with barely contained dissension. The opening decades of the nineteenth century had been difficult ones for political radicals and reformers. Anything that smacked of dissent seemed tainted by automatic association with the revolutionary French enemy across the Channel, and political opposition was firmly stamped upon. Spies and informers appeared to be everywhere; they even followed poets like Coleridge and Wordsworth around. Whilst Coleridge spent 1797 in the rural idyll of the Quantock Hills, he was under constant observation from a Home Office lackey who wrote diligent notes to his superiors back in London on the activities of the 'mischievous gang of disaffected Englishmen' and 'violent Democrats' under his surveillance.[1] The Seditious Meetings Act was designed to make any kind of mass political gathering illegal, though it caught plenty of other associations in its nets as well. The act even managed to catch out the earnest and far from revolutionary young Michael Faraday when meetings of the City Philosophical Society, of which he was a member, were briefly banned as subversive.[2] Any public gathering became suspect. Habeas corpus was suspended, which meant that those suspected of subversive activity and treasonous talk could be incarcerated without trial. Serious disturbances, such as the Littleport riots in 1816 and the Peterloo massacre in 1819, seemed to embattled authorities to be clear indications of troubled times.

These were dangerous times to hold unorthodox opinions. Opposition rumbled on, however, notwithstanding all the efforts at suppression. In bookshops, coffee-houses and taverns, conspirators conspired and seditious pamphlets circulated under the counter. Coleridge's friend and fellow traveller, John Thelwall, fulminated against those 'hireling plunderers' who had set out to 'declare open, inveterate, irreconcilable war … not only against the lives, properties, and liberties, but against, the opinions, feelings, inclinations' of the common man.[3] He had his sights set on Edmund Burke and his vituperative onslaughts on the leaders of the French Revolution and their supporters in England amongst radicals and natural philosophers. Burke had said that the Revolution's leaders were like 'geometricians' and 'chemists' who 'consider men in their experiments, no more than they do mice in an air pump, or in a recipient of mephitick gas'.[4] In amongst wild talk about the rights of man, some of those pamphlets talked about electricity as well. Burke thought of the Revolution as electric too, and its supporters the 'true conductors of contagion to every country'.[5] If electricity really was the stuff of life as radicals thought, then there was no soul, only organised matter. Without souls there was no need for the Church and without the Church to underpin it, what was left of the English constitution?

According to the radicals' philosophy, society ought to be governed by natural law alone, with no need for God, kings or aristocracy. Nature made everyone equal and that should be the governing principle of society as well. Amongst the new ideas leaking over from France was the notion that human life could be explained entirely without recourse to supernatural causes. The materialist philosopher Julien Offrey de la Mettrie had been hounded out of *ancien regime* France for maintaining that man was nothing more than a machine that could wind itself up. A succession of French natural philosophers – Geoffroy St Hilaire and Lamarck – promulgated theories of human development that linked men to animals and suggested there was no need for a vital principle to explain how life appeared. Life was simply the result of material organisation and that organisation could be explained quite straightfor-wardly as the outcome of the operations of natural law. Nothing else was going on. Lamarck had suggested that life began with spontaneous generation (sparked into being by electricity, of course) and developed through progressive evolution from simple to more complex organisms. Nature, by this account, was hard-wired for progress and there was no need to invoke some mysterious higher cause.[6]

This kind of filthy French atheism looked like very dangerous talk indeed to many in Britain. To impressionable young radicals, on the other hand, it seemed tremendously seductive. Increasing numbers of medical students, besotted with these new French theories about evolution and transmutation and easy prey for subversive rabble-rousers, were turning materialist. Following Waterloo and the res-toration of peace across Europe, English medical students flocked to Paris in search of education and enlightenment. French medicine was celebrated for its ration-ality and Parisian medical schools were famous for the quality of their teaching. They certainly appeared to be an attractive proposition compared with London's crowded lecture theatres. The English medical marketplace was looking increas-ingly crowded too, and anything that made a budding practitioner stand out from the herd looked like a good investment. But there was nothing sanitary – to Tory eyes at least – about the ideas these Parisian schools peddled. The students returned infected with Lamarckian nonsense. They believed in progress and the heresy that one species could be transformed into another. Worse still, they turned the materi-alist and transformationist notions they had acquired in French anatomical theatres into cudgels with which to beat the complacent and entrenched corporations that governed English medicine.[7] It was time to stop the rot.

That, at least, is what John Abernethy thought as he stepped into the breach to deliver his introductory anatomical lectures to the Royal College of Surgeons in 1814. Abernethy was very much an establishment figure. By the time he was appointed as Professor of Anatomy and Surgery to the Royal College of Surgeons in 1814, he had been surgeon at St Bartholomew's Hospital for almost thirty years. He was a respected writer on anatomical matters and had been elected a fellow of the Royal Society in 1796. He was particularly popular as a medical lecturer, offering private lectures to medical students in his own home. Indeed, these were so successful that the governors of St Bartholomew's Hospital built a lecture theatre in the hospital to accommodate the huge numbers flocking to his classes. As well as being a popular teacher, Abernethy was a fashionable and successful practitioner of his art, maintaining an extensive and profitable private practice, with a long list of wealthy and influential patrons.[8] Abernethy was a great admirer of the surgeon John Hunter, whose anatomical collections formed the basis of the Royal College of Surgeon's Hunterian museum, and saw it as his duty to uphold what he regarded as Hunter's views about the origins of life. Abernethy had a reputation for tough talking and plain speaking too. Now, as he faced up to the materialist challenge, he did not intend to mince his words.

The College of Surgeons formed the ideal backdrop for Abernethy's tirade against the radicals. Along with their counterparts at the Royal College of Physicians, they ruled English medicine. The college had its origins in the Company of Barber-Surgeons, founded in 1540. As the status of surgery rose during the eighteenth century, they parted company with the barbers to form the Company of Surgeons in 1745. In 1800, a few years after they left their medieval hall in the old City of London for plush new buildings at Lincoln's Inn Fields, they received their royal charter to become the Royal College of Surgeons in London.[9] The college president and its coterie of fellows wielded a great deal of power. The college had the right to license the practice of surgery within London, and becoming a member was to all intents and purposes a prerequisite of a successful medical career. To the rising generation of radical young medics, the men who governed the Council of the Royal College were the enemies of progress. Thomas Wakley, firebrand for reform and founding editor of the *Lancet* (established to cut out the corruption at the heart of English medicine), condemned them all as 'avaricious, cowardly, plundering, rapacious, soul-betraying, dirty-minded BATS'.[10] Abernethy, as one of the Royal College's founding fellows, was in radical eyes quite certainly a representative of the bats.

When he stepped up to the lectern at the Royal College of Surgeons, Abernethy fully intended to nail the lie that life was just a matter of organisation. It simply made no sense, he said, to suppose that life could just appear out of brute matter without some other agency being at work. 'In surveying the great chain of living beings,' argued Abernethy, 'we find life connected with a vast variety of organisation, yet exercising the same functions in each; a circumstance from which we may I think naturally conclude, that life does not depend on organisation.'[11] If life was not the result of material organisation, then clearly something had to be added to brute matter to endow it with vitality. In this, Abernethy put himself forward as defending the views of the venerable Hunter. According to him, it was 'Mr. Hunter's opinion, that irritability is the effect of some subtile, mobile, invisible substance, superadded to the evident structure of muscles, or other forms of vegetable and animal matter, as magnetism is to iron, and as electricity is to various substances with which it may be connected'.[12] Irritability, along with sensation, was widely agreed by eighteenth- and early nineteenth-century physiologists to be one of the two defining characteristics of living tissue.[13]

It was an old question: just what made living matter alive? And how did the soul communicate with the body it inhabited? Seventeenth-century thinkers such as Thomas Willis argued that the nerves were filled with animal spirits that mediated between the active soul and the inanimate brute matter of the body.[14] The eighteenth-century physiologist Albrecht von Haller placed the two characteristics of irritability and sensation at the centre of his definitions of life. He asserted:

> I call that part of the human body irritable, which becomes shorter upon being touched … I call that a sensible part of the human body, which upon being touched transmits the impression of it to the soul; and in brutes, in whom the existence of a soul is not so clear, I call those parts sensible, the Irritation of which occasions evident signs of pain and disquiet in the animal.[15]

That left the question of transmission open; how did flesh convey its messages to the soul? The English philosopher David Hartley suggested that the nerves were like wires that conveyed their messages by vibration – a thoroughly Newtonian answer to the problem, or so Hartley thought. But these were pragmatic solutions that dealt with what could be observed, and quite deliberately avoided the vexed question of life's source.

Matter, by the most common view, was just matter – cold, dead and inert. To make it live, it needed something more. This, according to Abernethy, was where electricity came in. In a bold flanking manoeuvre, he turned one of the radicals' main weapons against them. Life, he argued, was just like electricity. The 'phenomena of electricity and of life correspond', he averred.[16] In just the same way that electricity could be present in a wire without changing that wire's physical characteristics, so life could be present in matter without changing the material body's physical structure. He pointed to a whole series of analogies between electricity and the principle of irritability. Both were characterised by 'celerity and force'; both were vibratory. Humphry Davy's experiments, said Abernethy, had demonstrated that electricity was everywhere in nature. He argued, when 'therefore we perceive in the universe at large, a cause of rapid and powerful motions of masses of inert matter, may we not naturally conclude that the inert molecules of vegetable and animal matter, may be made to move in a similar manner, by a similar cause?' Davy's experiments and speculations on galvanism had turned fancy into fact. They had demonstrated 'the existence of a subtile, active, vital principle, pervading all nature as has heretofore been surmized, and denominated the Anima Mundi. The opinions which in former times were a justifiable hypothesis, seem to me now to be converted into a rational theory.'[17]

The difference between Abernethy's view of electricity, as opposed to that advocated by his radical foes, was that where they saw electricity as an intrinsic property of matter, he regarded it as something superadded to ordinary matter. He was by no means alone in his opinion. As we have seen, this was what Aldini's supporter the surgeon Charles Wilkinson had hinted at as well. Electricity was a vital principle that animated the entire universe, not just living matter. This, as far as the radicals were concerned, was nonsense – and it was not long before one of them stood up to say so. And what made the counter-attack even more painful for Abernethy was that, when it came, it was from his former friend and protégé. William Lawrence had started his medical career as John Abernethy's apprentice, living, as apprentices usually did, in his master's house. Abernethy had made him a particular favourite, giving him a job as his anatomical demonstrator at St Bartholomew's Hospital in 1800 – a position he held for twelve years. He soon made a name for himself as a skilled dissector and prolific author and in 1813, barely 30 years of age, became a fellow of the Royal Society. In 1815 Lawrence had been appointed Professor of Anatomy and Surgery at the Royal College of Surgeons, following in the footsteps of his old friend and patron John Abernethy. There was nothing particularly friendly about what he had to say about his erstwhile master's theories, however, when he in turn stepped up to the lectern to air his views before the college.[18]

As far as Lawrence was concerned, Abernethy's suggestion that the only way of explaining life was to invoke some superadded principle was beneath contempt. His robust counterblast poured cold water over the whole idea of a vital principle. Life, Lawrence thundered, was just a matter of organisation and that was all there was to it. As he explained to his audience:

> Life is the assemblage of all the functions, and the general result of their exercise. Thus organisation, vital properties, function, and life are expressions related to each other; in which organisation is the instrument, vital properties the acting power, function the mode of action, and life the result.[19]

Life, according to this point of view, was just one thing leading to another. The arguments put forward by Abernethy and his like were, according to the bellicose Lawrence, just so much hot air. They had no real idea what the vital principle they babbled on about was like at all. 'To make the matter more intelligible,' he scoffed, 'this vital principle is compared to magnetism, to electricity, and to galvanism; or it is roundly stated to be oxygen. "Tis like a camel, or like a whale, or like what you please"', he quoted from *Hamlet*.[20] If the vital principle was really like all these things, implied Lawrence, then it was like nothing on earth at all.

Abernethy's arguments were hinged in analogy. The vital principle was like electricity because it behaved in the same sorts of ways. Lawrence had no time at all either for analogical reasoning or for this analogy in particular. 'The truth is,' he claimed:

> there is no resemblance, no analogy between electricity and life: the two orders of phenomena are completely distinct; they are incommensurable. Electricity illustrates life no more than life illustrates electricity. We might as well say that an electrical machine operates by means of a vital fluid, as that the nerves and muscles of an animal perform sensation and contraction by virtue of an electric fluid.[21]

The real fact of the matter was that this farrago was just mysticism and rank superstition in another guise. Lawrence poked fun at his former mentor, comparing him mischievously to the 'Poor Indian, whose untutored mind/ Sees God in clouds, and hears him in the wind'.[22] These views might look like 'harmless reveries' to some, he warned, in the same vein as long discredited beliefs in sorcery and witchcraft. But when they were dressed up in the language of philosophical deduction, the charlatans who espoused them needed a sharp dressing down and a firm reminder that only the solid facts of observation and experience had a proper place in scientific investigation.

So far, so good. Lawrence's riposte may have dripped pure venom, but at least he had managed to avoid naming names, though there can be no doubt whatsoever that his listeners and readers knew exactly who he had in mind as his target. It was obvious that Abernethy knew as well. The tensions between the two erstwhile friends and colleagues simmered beneath the surface for a year or two, before exploding in very public confrontation in 1817. The arena, once more, was the Royal College of Surgeons, with Abernethy opening the skirmishing. In choosing to advocate and defend 'Mr. Hunter's theory of life', he told his audience he knew very well that he was asking for trouble. There existed a 'formidable Party' at large, he warned, who were out to subvert decent subjects of the Crown with sophistry and flashy rhetoric. These men possessed 'extensive information', were 'subtle disputants' and had 'words at will to make the worse appear the better argument'. Most damningly of all, they were 'writers by profession'. This was the party of 'Modern Sceptics'.[23] Abernethy was determined to set himself up against this bunch of shifty ne'er-do-wells, he told his listeners. Their arguments against the vital principle were mere flimflam and their 'crude speculations' brought the entire medical profession into disrepute. Now the politics was really out in the open.

This was more than enough to get the thin-skinned Lawrence going. Abernethy had tried to dampen the fire by avoiding naming names once more, but that did nothing to stop his adversary from going up in flames at this attack – as he saw it – on his integrity as a doctor and a gentleman. Up in front of the Royal College of Surgeons again in 1819, he defended himself vigorously against the charge that he

had 'perverted the honourable office, intrusted to me by this Court' by 'propagating opinions detrimental to society, and of endeavouring to enforce them for the
purpose of loosening those restraints, on which the welfare of mankind depends'.[24]
He scoffed at the notion that there was a party of 'modern sceptics' out there, intent
on bringing down the foundations of political society: 'A party of modern sceptics!
– A sceptic is one who doubts; – and if this party includes those who doubt, – or
rather who do not doubt at all, – about the electro-chemical doctrine of life, I can
have no objection to belong to so numerous and respectable a body.'[25] The details
of the pernicious doctrine were scarcely worth addressing: 'Living bodies, as well
as all dead ones, exhibit electrical phenomena under certain circumstances; but the
contrast between the animal functions and electrical operations is so obvious and
forcible, that the attempts to assimilate them do not demand further notice.'[26]

Now that things had become personal, Abernethy was not going to miss out on
the opportunity of putting the boot in either. Lawrence had always been an awkward bugger, he told his audience. Pulling rank as his upstart adversary's elder and
better, he regaled his audience at the Royal College of Surgeons with reminiscences
of how Lawrence, 'from a very early period of his professional studies … was accus-
tomed to decry and scoff at what I taught as the opinion of Mr. Hunter respecting
life and its function'. Lawrence, in other words, was simply up to his old tricks
again. He had been a cocky and disruptive student who had obliged Abernethy to
struggle 'in the midst of the controversy and derision of those students who had
become his proselytes'.[27] A cheeky upstart then, he remained a cheeky upstart still;
Abernethy implied he used his glib tongue and smooth way with words to poke
fun at his betters and cock a snook at authority. The difference now, of course, was
that he was involved in a far more dangerous game. Lawrence was making the
mistake of treating the strife-torn political battleground of post-war Britain as if it
were a school playground. It was one thing to raise fellow students in mock rebel-
lion against their schoolmaster, but quite another to use his philosophy as a weapon
to strike at the very roots of legitimate political authority.

By 1819 the increasingly vitriol-laden exchanges between Abernethy and
Lawrence in the Royal College of Surgeons' lecture halls had arrived at their bitter
climax. There were no further direct exchanges of fire between the battle-weary
main combatants. If the adversaries were at a loss for further invective, however, it
soon became clear that the theatre of operations had expanded well beyond the surgeons' headquarters at Lincoln's Inn Fields. Lawrence suddenly discovered that he
had become an icon for radicals and a tempting target for reactionary criticism. To
Tory minds, 'surgeon Lawrence' was to be lumped together with 'infidel Paine' and
Lord Byron in a 'radical triumvirate … colleaguing with the Patriotic Radicals to
Emancipate Mankind from all Laws Human and Divine'.[28] One of the first leading
the charge against the presumptuous bone-cutter was Thomas Rennell, appointed
a few years previously as Christian advocate at the University of Cambridge, and
charged with protecting delicate young undergraduate minds from just this kind

of atheistical claptrap. Like Abernethy, Rennell detected a 'fashion in scepticism' in the air, a sad symptom of the 'peculiar character of the day'. Nothing was safe from this sort of snide subversion:'no principle … either of natural or revealed religion, no one evidence, no one doctrine, which has not, in its turn, been captiously questioned, or rudely assailed.'[29]

As far as Rennell was concerned, this was all-out war, and he laboured under no illusions as to who the real enemy was and what they wanted. In his sights were 'the infidel school in France, and their copyists in Great Britain'. These men were out to 'destroy the relationship of the creature to the Creator, and to establish the independence of man upon God'.[30] A member of the influential Hackney Phalanx and editor of the *British Critic*, Rennell was just the kind of staunchly Tory high churchman that got up radicals' noses. His *Remarks on Scepticism*, published in 1819, was a well-timed counterblast to the pernicious materialism he saw flourishing around him. Lawrence and his like were out to 'impress upon the mind an erroneous notion of LIFE, and to represent it as entirely dependent upon organization for its continuance'.[31] Rennell thought this kind of thinking was not only immoral, but wrong headed. Lawrence was leading his followers down a garden path to ruin without even realising what he was doing. His 1819 essay, along with similarly-minded sermons from Cambridge pulpits, was an effort to pour cold water over what he regarded as the rantings of radical hotheads. The kind of materialism he read in Lawrence was an 'infection' that was 'subversive of all private happiness, all social morality'.[32]

Rennell was far from being the only one who thought like this. Other pamphlets ranted at Lawrence for having 'contrived to destroy the value of almost every other science'. The filth he peddled, they raged, was 'the doctrine of materialism in its grossest and most disgusting form'.[33] The invective was not all on one side however. Wylke Edwinsford was scathing about Rennell's writings: 'disgusted at the narrow spirit of intolerance and bigotry which pervaded his pages and sullied his clerical character.'[34] He accused Rennell of arguing like a schoolboy – taking quotations out of context and pouncing on verbal infelicities – rather than engaging seriously with the matter at hand. There were no holds barred in these pamphlet wars of words and ridicule was often the main weapon on both sides. Wylke Edwinsford in turn poked fun at poor Abernethy, dismissing his vital principle as 'a sort of "will o' the wisp," dancing about the body'.[35] Both sides were determined to portray their enemies as charlatans, unfit for the offices they held and with views unworthy of rational men. This was less argument than jockeying for position. Who one's allies were mattered more than just exactly what they said, as the opposing sides howled at each other across the barricades. The heat they generated was a fair indication of the passion that physiology could produce in Britain after the French Revolution. Arguments about the way the body was organised and how it functioned were arguments about the body-politic and the proper organisation of state and society as well.

When the *Quarterly Review* weighed into the argument they made it quite clear to their readers what they thought was at stake. This was a battle against the 'school of modern French philosophy' and its 'worn-out but mischievous opinions'. Lawrence was guilty of reducing 'man, the highest of the animal creation, with all his faculties of invention, memory, imagination', to 'an oyster or a cabbage'. In doing so, he was guilty of trying to turn the Royal College of Surgeons into 'a nursery for scepticism in religion, or republicanism in politics'.[36] The *British Review* was similarly contemptuous: 'It formerly suited the purpose of infidelity to spiritualise matter,' it scoffed, alluding presumably to Joseph Priestley's views on active matter, 'and now it seeks to gain its object by materialising spirit!' The belief that mind was different from matter and that the 'higher intellectual and moral powers of our nature' did not 'perish with the perishing body', was 'an inborn and spontaneous aspiration of the heart'. Its overthrow would be 'fatal to civilisation and happiness'.[37] Interestingly, the reviewer was clearly convinced that Wylke Edwinsford was a pseudonym, presumably for the publisher Richard Carlile.[38] The Unitarian *Monthly Repository of Theology and General Literature*, on the other hand, refused to 'stand timidly by and witness the scandalous opinions imputed to the Materialists, as consequences of their doctrine, and repeated in a geometrical progressive ratio with the solemnity and repetition of denial'.[39]

The link between body and polity that worried and disgusted the *Quarterly* and the *British Review* was certainly what got Coleridge interested in the debate. Coleridge – like Humphry Davy and other radical fellow travellers – had come a long way since the heady days of the 1790s. Unlike Davy, however, he still retained his passion for galvanism, and the use he made of it at this point shows just how ambiguous the versatile fluid could be in the changing political climate of Britain in the aftermath of the Napoleonic Wars. A hopeless opium addict, by the end of the 1810s Coleridge was a permanent lodger with his doctor, James Gillman, at his Highgate residence. He had long since abandoned the youthful radicalism that first piqued his interest in galvanic experiments. When he stepped into the debate with his essay, *The Theory of Life*, it was firmly on Abernethy's side of the barricade. In Abernethy's theories, he claimed, 'it is impossible not to see a presentiment of a great truth'.[40] The problem with modern philosophy, according to Coleridge, was that it had become synonymous with materialism. Mechanism was 'the sole portal through which truth was permitted to enter'. The result was that even the human body was treated like a hydraulic machine. Galvanism changed all this though: 'in every sense of the word, both playful and serious, both for good and for evil, it may be affirmed to have electrified the whole frame of natural philosophy.'[41]

Galvanism changed things, as far as Coleridge was concerned, because it provided a means of doing away with old distinctions between matter and spirit. Galvanism was both and neither. It was too subtle to be matter and too gross to be spirit. Galvanism as encountered in the material world was just an instance of a more general principle. It was one of a series of principles that bridged the gap between matter and spirit. Whilst Coleridge was dismissive of the notion that it made any sense at all to talk about material organisation as the cause of life, he was just as dismissive of the notion that it could be caused by some superadded fluid or ether: 'to whatever quintessential thinness they may be treble distilled, and (as it were) super-substantiated.' Life was an act, he said, a process, and the power of irritability in the living body was the equivalent of the power of electricity in an inanimate body. It was not life itself, but one of the 'constituent forms of life in the human living body'. What Coleridge was trying to suggest was that life was a thriving, directed process and that galvanism was one of the principles through which that process manifested itself.[42]

Coleridge, like Abernethy, was trying to turn materialists' own weapons against them. Materialists and radicals wanted to argue that the link between galvanism and the body, demonstrated by Aldini and his like, was proof positive of the mate-rial basis of life. Abernethy and Coleridge argued that it was just the opposite. Galvanism's place in the body showed there was more than just material agency going on. Heavily influenced as he was by his reading of German philosophy, Coleridge was trying to argue that far from showing that life was just down to organisation, the presence of galvanism in the body actually showed that the whole universe shared some of the properties of living things. Vitality, therefore, could not be reduced to matter because even ordinary matter had a degree of vitality of its own. Lawrence, faced with arguments like these and striving to hold his own in the materialist corner, found himself in the rather peculiar position (for a materialist) of trying to argue that electricity had nothing significant to do with the body at all; that there was nothing special about the presence of electricity in the body; and that it certainly had nothing to do with vitality. Vitality itself was just a chimera – an appearance produced by the organisation of matter.

A few years later, Lawrence suffered the final ignominy when he tried to pros-ecute publishers who had pirated his lectures. The Lord Chancellor declared that since his writings were blasphemous they could not be offered legal protection. The judgement was an explicit affirmation of the fact that making life material went beyond the pale in Regency Britain. It broke the chain that linked the human to the divine, and doing that added up to denying divine approbation of the entrenched social order. It was tantamount to denying the glue that held moral life together. Electricity was therefore dangerous because it looked as if it might just turn out to be the stuff that would dissolve those moral ties. There was a very good reason for the violent language that materialism's opponents spewed out; they really did think that these ideas were obscene. But to orthodox Tories and Anglicans, Coleridge's alternative looked none too pleasing either. To suspicious eyes, it looked very much like pantheism – the view that the entire universe was alive and endowed with divine power. Ideas like that simply muddied the waters of moderate and measured faith as far as the conservative and orthodox were concerned. Whichever way one looked at it, galvanism appeared to be extremely hazardous material.

FIVE

DISSECTING TOM WEEMS

So where does all this leave Tom Weems? Why was his body taken down from the scaffold and paraded through the streets of Cambridge that Friday in August 1819 before being meticulously and electrically dissected? It was clearly no coincidence that all this took place when it did. Galvanism mattered that summer; the bitter debate between John Abernethy and William Lawrence was at its most rancorous. The whole question of the proper relationship between the material organisation of the body and the origins of life had ramifications that went well beyond the merely academic. In the uncertain aftermath of the war with France, and with radical sentiment on the rise in the wake of increasing economic distress, ideas about the organisation of the body were seen to carry important messages about the organisation of the state. It seems clear that the electrical experiments carried out on Weems' body were intended as an explicit intervention in this bitter and highly charged debate. No one in 1819 could have carried out such a series of experiments without being painfully aware of the political dangers of galvanism. Mary Shelley's *Frankenstein* shows, if nothing else, how close to the surface of public consciousness the politics of electricity, life and death came at this time.

This was particularly true in Cambridge. The town and the county had their own recent reminders of political and economic volatility with the Littleport riots only a few years previously. Conservative university authorities were well aware that radicalism simmered amongst the student body. Thomas Rennell's pulpit thunderings, and his public vilification of William Lawrence and his like, were only one indication of the discomfort Cambridge felt at the erosion of old verities and the attraction dangerous new ideas held for some of its members. Weems' body, therefore, seemed to have arrived at an opportune moment for troubled Tories. This was their opportunity to rein in the radicals, reveal the relationship of electricity to the body as it really was and put the genie of galvanism firmly back where it belonged. Weems, as a result, was duly electrified, his body made to dance to the same galvanic tune as George Forster's a generation previously. The experimenters wanted to take their opportunity to put galvanism to the test – to show just what it could do and to what extent it could be seen to replicate the properties of life. At the very least, the experimenters were arming themselves with intimate knowledge of how electricity worked upon the body that they might turn to useful purpose. Their electrification of Weems' body was both an exercise in damage limitation and a preparation for combat.

One more element is needed to complete the picture, nevertheless. Aldini's experiments were not, as it turns out, the immediate inspiration for James Cumming's galvanisation of Weems. A very similar set of experiments had taken place only the previous year in Glasgow, carried out by another chemist, Andrew Ure, and their results had appeared very shocking indeed.[1] In 1818 when these experiments took place, Ure was a well-established figure in Glasgow. He came from a respectable family of city merchants and had studied medicine at Glasgow University before serving for a while as an army surgeon in the Highlands. In 1804 he had managed to secure for himself the Chair of Natural Philosophy in Anderson's Institution. This establishment had been founded only a few years earlier by the terms of the will of the recently deceased John Anderson, to teach mechanics and natural philosophy to local artisans. In many ways, it was a typical example of late Enlightenment patronage, aiming to produce social and moral improvement and progress through mental education. The position was unpaid, but Ure was soon earning a comfortable living from tuition fees. He prided himself on his success at ensuring that 'the treasures of natural philosophy are unlocked to the artisan'. Popular as Ure may have been as a lecturer (and the numbers flocking to his lectures certainly seems to suggest that), he was developing rather an unsavoury reputation too, as a rancorous, quarrelsome self-publicist with a keen and unfriendly eye for the main chance.[2]

A chance for fame was to all appearances what seemed to be on offer when Matthew Clydesdale came Ure's way. Clydesdale was in his mid-twenties in 1818, married with two children and a weaver by trade; he occasionally supplemented his living by working as a collier. His path crossed Ure's after he was found guilty at the Glasgow circuit court of murdering an old man, Alexander Love, earlier in the year. It appears to have been a pretty brutal murder, the victim having been savagely beaten around the head with a pick axe. In any case, the presiding judge, after sentencing Clydesdale to be hanged, passed the additional sentence that he should be 'delivered up by the Magistrates of Glasgow, or their officers, to Dr. James Jeffrey, Professor of Anatomy in the University of Glasgow, there to be publicly dissected and anatomised'.[3] The judge, Lord Gillies, made it clear to the convicted man that he should not waste time in entertaining any thought of the possibility of mercy. He would be hanged and dissected and that was all there was to it. Unsurprisingly, Clydesdale, along with some other prisoners condemned to hang at the same time, made a botched attempt at escape. According to one account, Clydesdale himself also attempted suicide, slashing his wrists with a bottle left him by one of the gaolers. Ironically, if the story is true, he was patched up in readiness for the gallows by the very man who would afterwards be responsible for his dissection.[4]

On Wednesday 4 November 1818 at two o'clock precisely, Clydesdale, along with one other condemned man, was hauled in front of the magistrates. The magistrates were informed by the Reverend Dr Lockhart that the prisoners had both confessed their guilt to their respective crimes (the other man, Simon Ross, had been convicted of housebreaking and theft) and the assembled company knelt together in prayer. Shortly afterwards, at a quarter to three, the crowded hall was cleared and the two men led out to the scaffold. Twenty minutes later they were both hanged. The crowd at the execution was substantial and a detachment of troops from the 40th Regiment of Foot, along with the First Dragoon Guards, had been told to guard the scaffold against any interference by the mob. Once Clydesdale's body had been cut down, it was placed in a coffin ready to be carted across the town to its final destination in the university's anatomical theatre. Much of the crowd (and the soldiery) followed. The cart was preceded by 'eight or ten of the town offic ers, in their black beavers, red coats, blue breeches, and white stockings with their halberts [sic] and battle-axes towering several feet above their heads'. As the procession made its way up the Saltmarket and along the High Street to the college gates, every window and roof was crowded with eager spectators jostling for a view.[5]

Mere dissection was not all that waited for poor Clydesdale's body once the college gates had shut behind him with a final knell, however. He was going to be galvanised – and Andrew Ure would be the man wielding the galvanic pile. Ure's unsavoury reputation seemed to have achieved new heights by 1818 and he almost seemed to some people to be going out of his way to cultivate notoriety. Earlier in the year, staid and presbyterian Glasgow had been rocked by scandal when it emerged that Ure's wife was expecting another man's child. Never the forgiving sort, Ure promptly and successfully sued the courts for permission to divorce his wife. The other man in the triangle was Granville Sharp Pattison, Ure's colleague and Professor of Anatomy at the Andersonian Institute. Pattison had a dubious name for himself too, and a reputation for dealing with grave-robbers to acquire cadavers for his anatomy classes. He resigned his professorship and decamped to the United States before the Andersonian's managers had the chance to sack him, complaining loudly (and remarkably) that the Ures had colluded together to engineer his disgrace. The scandal surrounding Andrew Ure's personal affairs had barely had time to subside before he flung himself into the public gaze again with his spectacular electrical dissection of Matthew Clydesdale's corpse.

Before the body landed on the slab before him, Ure made sure he had everything in order. He had prepared a powerful battery of 270 pairs of plates, charged with a solution of nitro-sulphuric acid. With Clydesdale before him, a number of careful incisions were made in the dead man's body so that different nerves were revealed ready for electrification. One end of the battery was connected to the spinal marrow, another to the sciatic nerve, with the result that every 'muscle of the body was immediately agitated with convulsive movements, resembling a violent shuddering from cold'. When the end of the battery was touched to the cadaver's heel, 'the leg was thrown out with such violence, as nearly to overturn one of the assistants, who in vain attempted to prevent its extension'. Next, with the phrenic nerve exposed, one end of the battery was connected to the diaphragm and the other to the phrenic nerve in the neck. After a little modification to the experiment, the results were more than satisfactory: 'Full, nay, laborious breathing, instantly commenced. The chest heaved, and fell; the belly was protruded, and again collapsed, with the relaxing and retiring diaphragm.' As long as the battery remained connected, it appeared that the lifeless corpse would continue breathing.[6]

There was more to come. When incisions were made to the face and the battery's ends touched to the appropriate nerves, 'most extraordinary grimaces were exhibited every time that the electric discharges were made'. In the final experiment, when the 'electric power' was transmitted from the spinal marrow to the ulnar nerve (in the elbow), 'the fingers now moved nimbly, like those of a violin performer; an assistant, who tried to close the fist, found the hand to open forcibly, in spite of his efforts. When the one rod was applied to a slight incision in the tip of the fore-finger, the fist being previously clenched, that finger extended instantly; and from the convulsive agitation of the arm, he seemed to point to the different spectators, some of whom thought he had come to life.' Which, of course, was just what Ure wanted them to think. As he gave his lurid account of the whole performance to the Glasgow Literary Society a little over a month later, he boasted that had he been able to follow his original plan of attack, 'there is a probability that life might have been restored'. His experiments showed, he averred, that if 'that substitute for the nervous influence' (electricity), was passed properly along the nerves in the way he had demonstrated, there was every prospect that it would awaken the 'dormant faculties' and really raise the dead.[7]

Ure's own account of these experiments was deliberately and carefully calculated to be shocking. There was nothing understated in the way he described how, as electricity was applied to Clydesdale, 'every muscle in his countenance was simultaneously thrown into fearful action; rage, horror, despair, anguish, and ghastly smiles, united their hideous expression in the murderer's face'. The effect, he said, outdid 'the wildest representations of a Fuseli or a Kean'. His readers would have known exactly what he meant. The painter Henry Fuseli was famous for his graphic and grotesque representations of the body and his nightmarish, sexually charged visions of the horrifying and fantastic. Edmund Kean was celebrated as the greatest Shakespearean actor of the day, famous in particular for his ability to represent emotional intensity in playing characters both villainous and tragic. Ure wanted his audience left in no doubt that galvanism was doing more than just twitching the muscles in the murderer's face – it really was making Clydesdale live again. He complacently related how 'several of the spectators were forced to leave the apartment, and one gentleman fainted'.[8] The thought of a scene shocking enough to make even a hard-nosed Glaswegian medical man swoon must have been a sobering one.

If attention was what Ure wanted, then his experiment with Clydesdale was eminently successful. His own account of the affair duly appeared in the Royal

Institution's *Quarterly Journal of Science*, well-placed to reach a distinguished and influential readership. Clydesdale's trial and execution had been comprehensively covered in *The Times*, and the account had included a brief discussion of the galvanic experiments Ure had conducted. A few months later, *The Times* returned to the story, reprinting from *The Scotsman* a lurid account of the proceedings under the banner: 'Horrible Phenomena! – Galvanism.' This time, the journalist dwelled lingeringly on the ghoulish details, recalling how 'one gentleman actually fainted from terror or sickness' and describing how 'the corpse seemed to point to the different spectators, some of whom thought it had come to life!'[9] In fact, the account seems strikingly similar in its language to the one Ure himself had penned for the Royal Institution's journal. The story clearly improved with retelling, however. Recalling the events he had witnessed that day from the vantage point of the best part of a century later, one old duffer described how the corpse had 'stared, apparently in astonishment, around him' and 'made a feeble attempt as if to rise from the chair whereon he was seated'.[10]

If Ure's own account and the lurid stories in the press were to be credited at all, then this was truly nightmarish stuff for galvanism's enemies. This was exactly the kind of thing that the Cambridge experimenters were determined to prevent. Weems was going to help them show that Ure's account of what had happened in Glasgow the previous year was pure fantasy. Electricity did not and could not, after all, imitate or reproduce the characteristics of life. As the *Cambridge Chronicle* recorded, James Cumming had 'prepared his powerful galvanic battery ... with the intention of repeating some of the experiments lately described by Dr. Ure of Glasgow in the Journal of the Royal Institution'. Cumming's battery, as it happened, was slightly smaller than the one Ure had used on Weems, being made of only 220 pairs of plates rather than 270 pairs. Nevertheless, as the report made quite clear, the 'galvanic influence was applied in the same manner as at Glasgow, by connecting the ends of the battery with the nerves to be acted upon, and running one of the connecting wires over the ends of the plates in the last trough'. Not only that, but Cumming and his assistants went through exactly the same repertoire of expected bodily effects as had Ure in Glasgow. They tried to get Weems to breathe, to perform facial contortions and move his arms and legs by judicial application of galvanism in just the same way that Ure had described.[11]

There was, of course, a reason for this slavish repetition and meticulous attention to detail. Cumming wanted to show that Ure was talking nonsense. To do that, he had to make quite sure that he had gone through exactly the same procedures as his opponent had. There could be no possibility of allowing Ure or his supporters to retort in turn that they could hardly have expected the same results since they had not performed the same experiment, or had not performed it properly. At the outset of its account (which in all likelihood had been penned by Cumming himself) the *Cambridge Chronicle*'s account made clear that 'the result of the experiments upon the body of Weems ... was exactly what we have witnessed on the bodies of animals, and equalled what we should have been prepared to expect, had we not read the above-mentioned account by Dr. Ure'. Where Ure had produced heavy breathing, Cumming produced a movement 'similar to a slight shuddering from cold'. Where Ure had coaxed Clydesdale's face into extravagant contortions, all Cumming could get out of Weems was a 'convulsive twitching' that was 'not expressive of any of the mental affections of life'. As the *Chronicle* primly concluded:

> From the account given by Dr. Ure, they had been led to form, perhaps, too san-
> guine an expectation of the efficacy of galvanism in restoring suspended animation;
> it is therefore to be regretted, that, although undertaken with a battery at least
> equally powerful, and with every attention to ensure success, these experiments
> afford no confirmation of the hopes held out by those at Glasgow.[12]

And that, as Cumming doubtless piously hoped, was that. The blasphemous Scotsman had been put firmly in his place. Galvanism's connections to the body had been authoritatively demonstrated and delimited. The genie was back in the lamp where it belonged. Tom Weems had done his bit after all to uphold Cantabrigian sensibilities and maintain the proper order of things. If that really is what Cumming thought, then he was on a hiding to nothing. The Scots them-selves were certainly not disheartened by Cumming's strictures. A few years later, they even tried to do it all again, on the body of William Divan. Ure was not there this time (having troubles of his own) and James Jeffrey performed the role of master of ceremonies. The result, perhaps because of the lack of Ure's enlivening presence, was disappointing; with Jeffrey admitting that 'the operations had not turned out altogether as well as could have been wished'.[13] He blamed the noose, rather than any failures of his own or of galvanism's revivifying powers, speculat-ing that it had rended and destroyed the nerves so that the passage of galvanism through the body was disrupted. Neither Cumming's experiments on Weems, nor Jeffrey's efforts with Divan did anything to damage galvanism's reputation as the ultimate restorer of life in any case. If nothing else, Ure was a better self-publicist than either of them, and it was his version that stuck.

Tom Weems' fate does, however, tell us a great deal about the ways in which debates surrounding bodily electricity during the early decades of the nineteenth century reflected far wider concerns about the state of society. Electricity mat-tered to all the protagonists in these bitter disputes because it spoke to serious concerns; just what could natural philosophy reveal about the nature of the soul and the politics of progress in the aftermath of war? When the good and the great of Cambridge society crammed themselves into the Chemical Lecture Room on that August afternoon – and even though we cannot be sure, it is very tempting to suppose that Thomas Rennell, for one, must have been there – they knew what they were going to see and they knew why it mattered. Everyone there would have understood what this experiment was really all about. No one would have laboured under the misapprehension that they were there to witness a disinterested investigation. No doubt they all expected a good show, and some may have been disappointed not to have been presented with a spectacle as grotesque and impressive as the one Ure and Clydesdale had offered their audience in Glasgow the previous year. But frivolity aside, they all expected to see orthodoxy restored (unless a few subversives had managed to sneak in with the crowd), and that, if only for a little while, was what Weems and Cumming gave them. In the end, Weems' dissectors had another cut in mind. What they really wanted to do was divorce galvanism from its politics: they failed.

What happened in the very end to Tom Weems we do not know for certain. We can reasonably conclude that what little was left of him after Cumming's and Oakey's attentions was buried in some anonymous grave. And what about the rest of the protagonists in Weems' story? Following his experiments on Forster, Aldini returned to Bologna and later settled in Milan, where he became a state councillor and was given a knighthood by the Austrian emperor for his services. Galvanism, as it turned out, brought him only comparatively fleeting fame. During the 1820s he busied himself with plans for improved lighthouses and with schemes to protect buildings and people from fire. In 1830 he was awarded a prize by the French Academy of Sciences for his labours. He visited London at least once again, presenting his researches on fire prevention to the Royal Institution. His facility for performance was clearly undiminished as his assistant, suitably clad, bathed his head in flame for several minutes; 'We never heard applause greater in any scientific assembly than attended this experiment', remarked one journal.[14] When he died in 1834, aged 71, he had amassed a significant fortune and left some of it to establish a school for artisans in his native Bologna.

Ironically, given the fate of his opponent William Lawrence's lectures, Abernethy during the 1820s also found his lectures pirated in the pages of the *Lancet*. Abernethy's main claim to fame, nevertheless – and despite the notoriety he acquired through his disputes with Lawrence – had always been as a practitioner rather than a teacher or a polemicist. His plain speech and forthright manner, according to some critics at least, was an act cultivated to impress his clientele amongst the upper crust. His career continued to flourish until a few years before his death in 1831.

William Lawrence also continued to make a name for himself as a radical firebrand throughout the 1820s – or at least until 1828, when the council of the Royal College of Surgeons elected him to their number. As a member of the elite, Lawrence promptly abandoned his radical politics and spent the rest of his career staunchly defending entrenched privileges. In 1857 he was appointed surgeon general to Queen Victoria and just before his death in 1867 he was rewarded with a baronetcy. Early radicalism had clearly done no harm to his career, as some of his obituarists sourly noted.

Shortly after his engagement with Matthew Clydesdale, Andrew Ure was again embroiled in controversy, this time of a personal nature, as he divorced his wife following the exposure of her affair with his fellow professor Granville Sharpe Pattison at the Andersonian's Institution. Despite a reputation for high-handed behaviour, Ure became a fellow of the Royal Society in 1822. In 1830 he abandoned the institution and left Glasgow for London, where he established himself as a consulting chemist. A few years later he published the notorious *Philosophy of Manufactures* following an enthusiastic tour of northern manufacturing districts. The book was a love letter to the mechanised factory, extolling the virtues of the 'processes that may be employed, to give to portions of inert matter, precise movements resembling those of organised beings'. The factory, marvelled Ure, was 'a vast automaton, composed of various mechanical and intellectual organs, acting in uninterrupted concert for the production of a common object, all of them being subordinated to a self-regulated moving force'.[15] If his experiments on Clydesdale were an attempt to display that bodies could act like machines, then by now he was keen to demonstrate the opposite too. Karl Marx was in due course to caustically dismiss Ure as 'the Pindar of the automatic factory'.[16]

Humphry Davy continued to flourish, of course, at the Royal Institution. He acquired a baronetcy and married well, but would never quite escape the charge that he had risen above his proper station. His lectures at the Royal Institution continued to draw in the fashionable crowds, and despite the fact that much of his reputation as a chemist had been acquired by doing down French chemistry, Napoleon invited him to Paris in 1813 to be awarded a prize by the French Academy of Sciences. Following the death in 1820 of the autocratic Joseph Banks, who had presided over the Royal Society for almost half a century, Davy found himself hoisted into the presidency – an awkward compromise candidate between the reformist and reactionary camps. He was not a success and certainly did not enjoy the honour. His erstwhile apprentice, Michael Faraday, was by now making a name for himself and – in Davy's view, at least – also developing a propensity for having ideas above his station. Faraday's nomination to a fellowship of the Royal Society looked to Davy like a deliberate attempt by his enemies to embarrass his efforts to keep the Society from veering too much towards either hostile camp. Davy died, alone, in Geneva in 1829, two years after gratefully handing on the reins of the Royal Society to his successor.

By the end of the 1820s electricity was certainly in the process of becoming a rather more respectable science than it had been a few decades earlier. That had been one of Davy's achievements – and his protégé Faraday would continue the process into the century's second quarter. The intimate connection between electricity and life that had seemed to underpin Aldini's experiments on Forster, Ure's experiments on Clydesdale and even James Cumming's experiments on the remains of the unfortunate Tom Weems, seemed less self-evident than it had done a decade earlier. Galvanism still remained politically dangerous though. Aldini's and Ure's experiments – if not Cumming's efforts with Weems – could still be cited as evidence of materialism and support for the underpinnings of radical politics.

In fact, it was probably precisely because of the lack of a radical story about galvanic life that Cumming's experiment, to a large extent, simply sank without trace. It was Aldini's and Ure's flamboyant and suggestive performances that survived. However 'disgusting' Davy (and probably Faraday, too) thought those performances were, they played a crucial role in defining electricity in the opening decades of the nineteenth century for much of the interested public. The electricity those performances made concrete, including Weems' posthumous performance, was corporeal and sensational. It was a fluid intimately connected to the body, seductive and seditious in the messages it conveyed.

PART II

ADA LOVELACE

SIX

KNOWING ADA

This was not going to be an ordinary visit between neighbours, but the there was little ordinary about either Fyne Court or its inhabitants when Ada Lovelace came to stay in the late autumn of 1844. Ada Lovelace was famous then, just as she is now, as the poet Byron's only (legitimate) daughter. She was starting to become famous (or notorious) for her mathematical talents as well. Her background and her father's reputation meant that she could move quite easily in exalted scientific circles. She could consult Charles Babbage and Mary Somerville about her mathematical education, as well as be tutored by Augustus de Morgan, Professor of Mathematics at University College London. Babbage, by the time Ada knew him, was a leading light in London scientific society. He held the Lucasian professorship of mathematics at Cambridge, the chair once held by Isaac Newton, whilst Somerville was celebrated as the English translator of Pierre Simon Laplace's *Mécanique Céleste* in 1831, and was the author of *On the Connexions of the Physical Sciences*, published in 1834. Furthermore, her fellow Scot David Brewster called her 'the most extraordinary woman in Europe'. Ada Lovelace's connections with people like this underline the importance of family ties and social status in establishing a place for oneself in the world of late Hanoverian and early Victorian science.

Ada was visiting Fyne Court and its owner, Andrew Crosse, for a very specific reason. She wanted Crosse's help to pursue what she regarded as her life's vocation. She wanted to use electricity to learn about her own body – she was chronically ill – and to use her own body as a living laboratory for the investigation of electricity and its place in the universe. Ada had great ambitions to become a 'Newton for the Molecular Universe'[1] and to 'bequeath to the generations a Calculus of the Nervous System'.[2] All of this meant that she had to learn how to do something very unusual indeed (and possibly even rather disreputable) for a woman, even one of Ada's formidable reputation. She had to learn how to experiment, and she hoped that Andrew Crosse would be able to provide her with the necessary tuition. An interest in science was unusual enough for a woman – other than as a member of the audience.[3] A handful of women, like Somerville or her fellow author Jane Marcet, wrote successful scientific books, but very few got their hands dirty at the laboratory bench. Even Mary Somerville stopped short of that. Ada's ambition and her urgent need for a tutor in experiment were therefore highly unusual.

Crosse was not Ada's original choice, nevertheless. She had approached Michael Faraday first, only to receive a polite and tactful rebuff. The eccentric owner of Fyne Court was, however, well suited to the kind of experimental programme that Ada had in mind – considerably more suited than Faraday, as we shall see soon enough. Crosse was notorious in Somerset as the 'thunder-and-lightning-man'. 'You can't go near his cursed house at night without danger of your life,' the locals grumbled. 'Many folks have seen devils dancing on the wires he's had put round his place.'[4] Those neighbours with longer memories – and capable of recognising Crosse's garden adornments as wires for his experiments on atmospheric electricity rather than evidences of devil worship – would remember the Crosse family's long-standing radical credentials. Despite having acted as high sheriff of Somerset only a couple of years earlier, Andrew's father, Richard Crosse, prided himself on the fact that he had been in Paris during the Revolution and had – according to family legend at least – been one of the first to raise the tricolour on the ruins of the Bastille the day it was captured.[5] More recently, Crosse had achieved national notoriety when one of his experiments designed to produce artificial crystals by electricity, produced insects instead. Despite Crosse's denials, the experiment was hailed by radicals (and roundly condemned by conservatives) as evidence of electrical spontaneous generation. Unsurprisingly, it was one of the experiments that Ada wished to repeat.

Ada's father, George Gordon Byron, already famous as both poet and rake, married Anne Isabella Milbanke on the 2 January 1815. Annabella (as she was usually known) had done most of the running during their courtship. Byron had proposed to her once before but had been rejected. The second time round, it was Annabella who initiated correspondence and Byron who was half-hearted about the whole affair. The marriage was a disaster. Byron had married her mainly for money, and when it transpired that relatively little was to come his way from Annabella's debt-ridden father, he rapidly lost interest. Annabella was just as quickly disillusioned. Byron drank and had affairs. An only child whose every whim had been gratified by her adoring parents, Annabella had a very rude awakening as she realised just how little her desires figured in her husband's reckoning of things. 'Had Lady Byron on the sofa before dinner,' wrote Byron brutally in his diary on their wedding day.[6] Ada Augusta Byron was born on 10 December 1815. Within a month of the birth, Ada's parents had separated and Byron soon left England and never returned. Ada therefore grew up with no memory of ever having set eyes on her illustrious father.

If, on the one hand, Byron made his feelings for his little daughter quite clear by abandoning her, his poetry reveals rather more ambiguity. Several passages in *Childe Harold* were devoted to Ada. She was the 'sole daughter of my house and heart'. The third canto is redolent with regret for having lost his daughter to a life of wandering and doubt in his self-exile, thinking of her as:

> Soul of my thought! with whom I traverse earth,
> Invisible but gazing, as I glow,
> Mixed with thy spirit, blended with thy birth,
> And feeling still with thee in my crushed feelings' dearth.

This was all self-indulgent tosh, of course. Given the legal situation of husbands and wives with regard to their children at the time, Byron could certainly have kept Ada with him had he wanted to do so. In later life, Ada appears to have expressed a great deal of affection and pride in her absent (and by then romantically dead) father. At her own request, following her death her remains were laid next to his in Hucknall Torcard Church, near Newstead Abbey – the Byron family seat until the improvident lord sold it in 1818. One American admirer reported how she 'loves to talk of her father, and glows with delight when you tell her that his works are universally read, not only in the seaboard cities of America, but among the far-away woods and prairies of the New World'.[7]

Her mother, on the other hand, was determined that Ada would inherit nothing except the name from the man who had abandoned both of them, and she spent much of Ada's childhood worrying that she might take after Byron. Annabella was firmly resolved to 'endeavour seriously to consider and diligently to execute the duties of a Mother, and to divest myself of wrong bias arising from my particular circumstances or morbid feelings'. She thought that Ada needed 'watchful & judicious superintendence' to 'form the basis of good habits, & prevent the rise of evil ones'.[8] When Ada was 3, Annabella boasted that she was a 'happy and intelligent child, just beginning to learn her letters – I have given her this occupation, not so much for the sake of early acquirement, as to fix her attention, which from the activity of her imagination, is rather difficult'.[9] Unsurprisingly, Ada was anxious 'to please Mama very much, that she & I may be happy together'. Ironically, as a child, she recorded that 'I do not like arithmetic & to learn figures'.[10] One acquaintance recorded that Ada's 'feelings towards her mother were more akin to awe and admiration than love and affection'.[11] Annabella, in turn, was clearly terrified that her daughter would inherit her father's loose morals – a fear that seemed to be realised when the teenage Ada was caught out in an affair with her tutor. Sophia de Morgan, a family friend and wife of Ada's future mathematics tutor, recalled that there had been 'no *real* misconduct at that time … but it was very evident that the daughter who inherited many of her father's peculiarities also inherited his tendencies'.[12]

Like all mothers of her class, Annabella was determined that her daughter should marry – and marry well. Youthful sexual indiscretions that might be overlooked (or even discreetly applauded) in a man were unlikely to be so overlooked in a young woman. In any case, by the 1830s the strait-laced, middle-class and evangelical morality that would come to characterise the public face of the Victorians was already starting to infect the aristocratic classes. Annabella was therefore delighted when William King, eighth Baron of Ockham, proposed and was accepted by her wayward daughter. They were married on 8 July 1835. A few years later, William was elevated to become the first Earl of Lovelace and Ada accordingly became Lady Lovelace. Her husband was a graduate of Trinity College, Cambridge, and had made a career for himself as a diplomat and politician. Like both the Byron and Milbanke clans, William's family were firmly established in the liberal Whig tradition – a key consideration in forming important family alliances. Less importantly, as far as Annabella was concerned, William also shared many of Ada's interests in mathematics and the natural sciences. He was to become a fellow of the Royal Society in 1841.

By the time she and William were married, Ada was already convinced that her future lay in mathematics and the philosophy of nature. She was also very ill. Her friend Woronzow Grieg – Mary Somerville's son by her first marriage – remembered her from their first encounter as 'rather stout and inclined to be clumsy, without colour and in delicate health'.[13] By the mid-1830s she was a confirmed invalid, and as Woronzow Grieg recalled, she suffered frequent dizzy spells. In 1835, a few months before her marriage, she wrote to Mary Somerville to apologise for her ill health. She was, she said, 'for some unaccountable reason in a weak state … When I am weak, I am always so exceedingly terrified, at *nobody knows what*, that I can hardly help having an agitated look & manner, & this was the case when I left you'.[14] In 1837 following the birth of her daughter, also named Annabella, Ada suffered from what she and her physician, Charles Locock, suspected might be cholera. She worried about the equilibrium of her precarious physiology, telling her mother she thought she had over-eaten in response to the attack of cholera and 'that some starvation has become subsequently necessary, as a balance to my having over-loaded my system'.[15] Before her visit to Andrew Crosse, Ada warned him discreetly about her 'gastritis'.

Ada regarded mathematics as her salvation – bodily and mentally. She had first turned to science as a penance for her teenage indiscretions. 'I find that nothing but very *close & intense* application to subjects of a scientific nature now seems at all to keep my imagination from running wild,' she confessed to her tutor.[16] Her tutor agreed that such subjects were indeed a 'moral discipline, tending to control the imagination, and give one mental self command. I should recommend you a complete Cambridge course.'[17] Actual Cambridge study was, of course, beyond Ada's feminine reach, but she did her best to make up for it by deluging her then tutor, William King, (no relation to her future husband) with mathematical problems and requests for new texts to devour. She wanted to master 'Euclid, and Arithmetic & Algebra' as quickly as possible so that she could move on 'to make myself well acquainted with Astronomy, Optics &c.'.[18] She wrote to William Frend, the disgraced former Cambridge mathematician and radical sympathiser, asking for advice about rainbows. She was also soon in contact with Mary Somerville and writing to her for advice on mathematical and scientific matters. Somerville was happy to help since she had acquired much of her own knowledge of such matters in similar fashion. Ada was soon enthusiastically trying to spread the mathematical gospel herself as well, taking on her mother's god-daughter as a pupil and bullying the poor girl about mathematical self-discipline in much the same way that her tutors had bullied her.

By now, Ada had already met Charles Babbage – the man whose Calculating Engine would make her famous. She met Babbage during her first London season during the summer of 1833; Babbage's soirées were famous. As Andrew Crosse's wife remarked, recollecting London's scientific scene more than half a century later: 'One of three qualifications were necessary for those who sought to be invited – intellect, beauty, or rank.'[19] Even if Ada did not qualify on intellectual grounds (which she did), her looks and her pedigree as Byron's daughter would certainly have gained her entry. Babbage promptly whisked her off to admire his engine, albeit with her mother firmly in tow. Babbage had by now been working on his ambitious plans for a calculating engine for well over a decade, ever since he and John Herschel had fantasised during their undergraduate years at Cambridge about the possibilities of calculating by steam. By 1833, with radical plans to reform the hidebound Royal Society root and branch in tatters and the Calculating Engine's future looking increasingly unsure, Babbage was rapidly turning into a bitter man. Ada's enthusiasm must have been refreshing. They were soon exchanging mathematical letters and during the early 1840s embarked together on a joint endeavour to generate important new publicity for the Calculating Engine.

An Italian engineer, Luigi Menabrea (later to be Italian prime minister) had just published a short description of the latest version of Babbage's invention – the Analytical Engine. With Babbage's help, Ada undertook to produce a translation of the treatise into English, along with her own extensive interpretative notes. The Analytical Engine received its operating instructions through a series of punched cards, an idea that Babbage had derived from the mechanism of the recently invented Jacquard Loom. Ada expressed it pithily: 'the Analytical Engine weaves algebraical patterns just as the Jacquard loom weaves flowers and leaves.'[20] The resulting publication in Richard Taylor's *Scientific Memoirs* in 1843 was a great success. The article and notes were discreetly signed AAL and Ada basked in the congratulations of those who knew what, or rather who, those three letters stood for. She saw herself on the brink of a life in science and started feverishly planning her next publishing enterprise. She was already boasting to Babbage, even whilst the proofs of her contribution to Taylor's journal were still being corrected, that before 'ten years are over, the Devil's in it if I have not sucked out some life-blood from the mysteries of this universe, in a way that no purely mortal lips or brains could do'.[21] She planned that her next project would be a similar translation with notes of the latest electrical discoveries of the German natural philosopher Georg Simon Ohm.

It looked as if Ada had found her place in the scientific world. London, at the beginning of Victoria's reign, was the capital of European science. The city's elite natural philosophers – the gentlemen of science – congregated at the Royal Society or at one of the more recently established specialist scientific societies, such as the Astronomical Society or the Geological Society. These were closed shops, open only to their members. More accessible, to those like Ada with money and connections at least, were the Royal Institution and its rival, the London Institution. The Royal Institution's location on fashionable Albemarle Street advertised its aristocratic affiliations. The London Institution in Finsbury Circus near the city had a more commercial and utilitarian bent. Each summer, much of scientific London would pack its bags and depart for some provincial town or city – Manchester, Liverpool, York or Bristol, say – for the annual gathering of the British Association for the Advancement of Science. There, the gentlemen of science had a jamboree: delivering addresses, airing their views on the latest discoveries and controversies, politely squabbling and renewing philosophical acquaintances. There was a thriving informal intellectual circuit as well. Charles Babbage's parties were famous, but they were far from being the only ones.

The gentlemen of science had very firm views as to who was – and was not – fit to call themselves a man of science. Science was a vocation, a calling for those with disciplined minds. The specialist scientific societies had been established by these vocationally minded gentlemen as breakaway groups from the conservative and dilettante Royal Society. Throughout the 1820s, Babbage and his fellow firebrand John Herschel had battled the Royal Society's old guard for supremacy in the scientific corridors of power. Herschel was put up against the Duke of Sussex (brother of the king) for the presidency in 1830. The battle for reform and meritocracy failed then, but by the 1840s the capture of the Royal Society was on the reform agenda once more. Women certainly had their place in this world of science. Cornelia Crosse remarked that 'amongst the scientific men of that day, there was a marked respect for female intellect'.[22] But it was a respect that depended on the woman knowing her place. Ada Lovelace's friend and mentor Mary Somerville was extremely unusual in being taken seriously as a scientific author by her male contemporaries, and even she was mainly valued as a communicator rather than a producer of scientific knowledge. Ada would have to struggle hard to gain that kind of favoured status.

Andrew Crosse, like many provincial men of science, was on the margins of this polite and metropolitan scientific scene. When Ada Lovelace came to visit in 1844, Crosse was 60 years old. He was proud of his family's radical traditions and that his father had been friends of Enlightenment radicals and natural philosophers such as Benjamin Franklin and Joseph Priestley. Crosse described his own political stance as 'anti-parsonic and a liberal to the backbone'.[23] He had dabbled in electrical experimentation throughout his life. Humphry Davy, the chemist and ailing President of the Royal Society, had visited Fyne Court in 1827 in company with Thomas Poole, his and Crosse's mutual friend. He had been impressed with Crosse's laboratory. Through Poole, amongst others, Crosse had also been friendly with the poets Coleridge, Southey and Wordsworth during their radical early years. Fyne Court was only a few miles from Nether Stowey in Somerset's Quantock Hills, where the three poets lived for a time, penning poetry and planning radical utopian experiments in communal living. They certainly shared his enthusiasm for galvanic experimentation, and indeed Crosse's second wife, Cornelia, suggested that his own obsession with electricity 'owes much to the divine afflatus of poetry'. Crosse himself declared that 'metaphorically speaking, electricity is the right arm of the Almighty'.[24]

Crosse had turned Fyne Court and its grounds into one vast electrical laboratory. The wires strung from the trees – which some of his neighbours thought were tightropes for dancing devils – were in fact part of Crosse's apparatus for investigating atmospheric electricity. The wires were arranged around the woods of the estate and linked to a huge battery of fifty Leyden jars kept in the music room at Fyne Court. When conditions were good, the huge battery could be 'charged and discharged twenty times in a minute, making a noise like that of a brisk cannonade'. All it took was a 'dense fog, or the muffled silence of a snowstorm' to 'supply the storage force required for those manifestations of electric intensity'. Then, if 'the receiving balls in the organ gallery were left two or three inches apart, the flashing and cracking would often continue with playful intermittence throughout the livelong night'. The apparatus' sensitivity was demonstrated one day when, without apparent warning, 'a very strong explosion' took place between the receiving balls. After the first shock 'many more took place, until they became one uninterrupted stream of explosions, which died away and recommenced with the opposite electricity in equal violence'. This went on for more than five hours, whilst the 'stream of fire was too vivid to look at for any length of time'.[25]

Crosse had been carrying out experiments like these at Fyne Court for at least thirty years by the time of Ada's visit. His other great galvanic obsession was producing crystals by electricity; these experiments dated from 1807. His first experiment was carried out with water from nearby Holwell cavern and succeeded in producing 'some sparkling crystals upon the negative platinum wire, which proved to be carbonate of lime, attracted from the mineral water by the electric action'.[26] In the perpetual darkness of the 'crystal room', Crosse subjected different solutions to low electric currents from a water battery for weeks and months, patiently watching to see if crystals developed. It was an attempt to imitate 'the slow and silent processes of Nature' and prove that electricity really was the force 'which lines the fissures of earth with metallic lodes, and sets the form of the leaf, or the facets of the crystal'.[27] The apparatus doubled up as medical equipment, and Crosse offered electrical therapy to the local poor suffering from paralysis or rheumatism. Until the middle of the 1830s, all this work took place in 'perfect intellectual isolation'. As his wife noted, at this stage in his life, he 'had no scientific friends with whom to communicate'.[28] All this changed, however, some eight years before Ada's visit, when Crosse's slow and patient electrical cultivation of crystals bore fruit that ended up catapulting him from his rural idyll right into the scientific limelight.

In 1836, the British Association for the Advancement of Science (BAAS) visited Bristol for its annual gathering of scientific gentlemen. The BAAS (or British Ass, as it was also known) had only recently been established, having had its inaugural meeting at York in 1831.[29] It was already developing into a powerful forum that metropolitan, Cantabrigian and Oxonian natural philosophers could use to direct provincial science. It provided excellent opportunities for self-promotion on the part of the provincial crowd as well, giving them a stage on which to display themselves in front of the elite. It was to describe the participants in this annual scientific jamboree that William Whewell first coined the word 'scientist'.[30] The BAAS was a peripatetic institution, moving from provincial city to city each year – and never meeting in London. Meetings stretched over several days towards the end of the summer and audiences thronged there in their thousands. They were treated to a hectic schedule of scientific papers delivered to the various sections, as well as outings to local sites of particular scientific interest. Newspapers and the popular weeklies devoted acres of newsprint to describing the proceedings. Charles Dickens, in *The Pickwick Papers*, poked fun at it all as the Mudfog Association for the Advancement of Everything.

Crosse was not sure whether he should attend the Bristol gathering, and was only eventually persuaded by his old friend Thomas Poole that this was an opportunity to engage with gentlemen of science that he could not afford to miss. At a gathering of the Association's Geological Section, after listening to 'the manly eloquence of Dr. Buckland, the pure intelligence of Professors Sedgwick and Phillips, the elaborations of President Murchison, the shrewd reasoning of Mr. de la Beche',[31] Crosse found himself the centre of attention. He was dragged on to the stage by Buckland and introduced as 'a gentleman ... whose name he had never heard til yesterday; a man unconnected to any society, but possessing the true spirit of a philosopher'. The gentlemen of science wanted to hold Crosse up as an example of the disinterested and humble man of science, in search of nothing beyond nature's truth. Not to be outdone, Buckland's geological rival Adam Sedgwick stood up to laud Crosse's electrical discoveries, insisting to the humble inquirer of truth that 'from this time forth he must stand before the world as public property'. Crosse, as his friend Thomas Poole boasted, had become 'one of the great show-beasts of the meeting'.[32]

Before disappearing back to Fyne Court and Broomfield, Crosse with characteristic generosity invited any gentlemen of science who were heading his way to visit and inspect his laboratory. Some of them took him at his word. Sir Richard Phillips, editor of the *Philosophical Magazine*, was the first to arrive. He and Crosse had long discussions about electrical philosophy. Phillips rhapsodised about the 'universality of matter and motion in producing all material phenomena' and the wonders of 'that universal harmony which surprises beings who, in eternal time, live and observe within only an unit of time'.[33] He admired the arrays of batteries Crosse had accumulated for his experiments, which 'resembled battalions of soldiers in exact rank and file, and seemed innumerable'.[34] Cambridge's Professor of Geology, Adam Sedgwick, was the next to arrive. Both were pressed to stay for 'a dinner as well served as I ever saw in any state dinner in London' and were asked to stay the night. Phillips decamped the next morning as 'continual fresh arrivals rendered it ineligible for me to prolong my visit'.[35] Crosse found himself suddenly transformed into a scientific celebrity, held up by the gentlemen of science as a model of devoted and disciplined inquiry. This new-found celebrity made the events of the next few months even more scandalous than they would otherwise have been.

Late in 1836, a little after Crosse's triumphant entry into scientific prominence at Bristol, something more than just crystals emerged from one of his long-running electrical experiments. Crosse had been carrying out some experiments in which a solution of silicate of potassa in hydrochloric acid was dripped slowly on to a porous piece of volcanic stone (collected on Mount Vesuvius as it happens) through which a low electric current was being passed. Crosse's 'object, in submitting this fluid to a long-continued electrical action through the intervention of a porous stone, was to form, if possible, crystals of silica at one of the poles of the battery'. Something far stranger than that happened, however. Over a period of several weeks, Crosse observed 'a few small whitish excrescences or nipples projecting from about the middle of the electrified stone'. He took these to be incipient mineral formations and continued to observe until, 'on the twenty-sixth day each figure assumed the form of a perfect insect, standing erect on a few bristles which formed its tail'. A few days later, the insects could be seen to move their legs. A few days after that, they were freely moving around on the stone.[36] Crosse's first inclination was to keep these startling results to himself; he was not a prolific publisher in any case. News of the amazing production of life by electricity soon leaked out through a visitor, nevertheless, and Crosse was in the newspapers once more.

If it had been anyone but Crosse, fresh from his Bristol debut, he would have been pilloried – he came very close even as things were. There were even reports of a local vicar conducting an exorcism at Fyne Court to drive out the devil that had undoubtedly occupied the place. Crosse was roundly condemned for meddling with things beyond the purview of science, despite his protestations that the experiment had not been designed to produce life and that he had no view at all as to what had really happened in the experiment. He complained bitterly that he had 'met with so much virulence and abuse, so much calumny and misrepresentation, in consequence of these experiments, that it seems, in this nineteenth century, as if it were a crime to have made them'.[37] His new friends amongst the gentlemen of science stood by him, nonetheless, unwilling to abandon their hero from the Bristol meeting. Crosse sent specimens of his strange insects – now dubbed the *acarus crossii* – to London. Richard Owen, the Hunterian professor at the Royal College of Surgeons (hardly a bastion of radicalism), duly examined them and they were put on show at one of the Royal Institution's Friday evening discourses. Both Owen and Faraday were discreetly circumspect about their views concerning the insects' origins. The geologist William Buckland, anxious to preserve Crosse's reputation as an experimenter without committing to the heresy of electrical life, suggested that electricity had been responsible for re-vivifying fossil insect eggs lying dormant in the volcanic stone.[38]

Radicals, of course, were delighted with Crosse's experiment. It was to become one of the cornerstones of their claims concerning the material origins of life. Other electricians responded to the challenge by attempting to repeat the experiment. Two of them – William Henry Weekes, who shared Crosse's fascination with atmospheric electricity, and Alfred Smee – reported a measure of success in these efforts. Ada Lovelace was herself keen to attempt the experiment as well. One of Crosse's poetical acquaintances responded in his own way:

Sparkle beneath both setting sun
And rising? Yet all this is done:
Nay, more: another insect I
Quicken by electricity.
My friend the generous Crosse will own
Life giving is not his alone.

This was Crosse's old friend Walter Savage Landor, poking gentle fun at the troubled experimentalist.[39] Crosse withdrew, injured, from the scientific limelight. But he was not excommunicated from the gentlemen of science's circle. Fyne Court remained a convenient stopping point for natural philosophical travellers heading for the West Country, with Buckland, Sedgwick and William Whewell amongst the visitors. For Ada Lovelace, residing relatively nearby in Ashley Combe, Crosse was a potential tutor in the art of electrical experimentation. Sufficiently respectable and moving in the same exclusive scientific circles as herself, he might also be expected to have some sympathy for the grand vision of electricity she was anxious to explore.

Ada Lovelace arrived at Fyne Court around 20 November. Andrew Crosse had come to visit the Lovelaces at Ashley Combe first, and after a few days there the party proceeded together towards Broomfield. Coming down to breakfast the following

Saturday morning, Ada found 'no *symptoms* of either breakfast (or cloth being laid), or of human beings'. They had all been 'sat up reading & talking philosophy till *one o'clock* last night'. Ironically, the 'droll thing was that we were discussing the metaphysics of *Time & Space*; & in so doing we forgot *real* Time & Space'.[40] She complained light-heartedly that there was 'no *order* ... You cannot conceive what an *odd* house it is altogether'.[41] She was amused and embarrassed to find that 'the *Water-closet* can only be got at *thro' the Drawing Room*; & of course it is perfectly evident the errand one is going on, since the exit leads nowhere else'.[42] She found all the talk exhilarating, however, though she thought that Crosse's daughter and his second son Robert were 'particularly *little* agreeable' and had 'a *bar sinister* about them; – a something or other of a *deep & sinister reserve* which is withering'.[43] Maybe they were just shy of their father's illustrious guest. Ada discussed electricity with Crosse and German metaphysics and mathematics with Crosse's son John. By the end of her stay she was sure that her 'visit to Broomfield seems to have been very opportune & to have laid the foundations for much that is wanted for me just at this epoch of my progress'.[44]

Ada's ambition was a grand one. She was aiming high indeed, for anyone – and particularly for a woman, however eminent her connections. Her aim was of 'one day getting *cerebral* phenomena such that I can put them into mathematical equations; in short a *law* or *laws*, for the mutual actions of the molecules of *brain*; (equivalent to the *law of gravitation* for the *planetary & sideral* world.)' For this, she needed to become 'a most skilful *practical manipulator* in experimental tests; & that, on materials difficult to deal with; viz: the brain, blood, & nerves, of animals'.[45] This was why she needed Crosse's assistance. Crosse could teach her those tricks of experiment that could not be mastered by reading scientific papers – how to manage a battery, for example, to keep the current constant, or how to wind coils of wire to achieve the maximum effect. With Faraday unwilling to help, Crosse, with his experiences of electrically induced life, must have seemed an ideal potential tutor. She assured him that in her view the 'intellectual, the moral, the religious' were 'all naturally bound up and interlinked together in one great and harmonious whole', and that the 'physical and the moral facts of the universe' were 'naturally related and interconnected'.[46] Her 'Calculus of the Nervous System' would be an attempt to unravel these connections.

Ada's position in 1844 is fascinating since she sat, in many ways, on some of the major fault lines of early Victorian scientific culture. On the one hand she was clearly part of the gentlemen of science's circle. As a result of her family connections and her social place, she rubbed shoulders with Charles Babbage and corresponded with Michael Faraday. She attended Babbage's soirées and Friday evening discourses at the Royal Institution. She had relatively easy access to the otherwise closed world of the scientific elite, simply because of who she was and who she knew. At the same time, her interests moved her in other directions as well. She took seriously ideas and phenomena that many gentlemen of science regarded as risible, radical or dangerous. Her interest in bodily electricity was certainly a concern that would have had little or no resonance (other than negative ones) for the men of science with whom she had most contact. Following Ada around therefore provides a way of exploring the complex and contested terrain of early Victorian scientific culture. The next two chapters will provide us with an overview of the electric world that Ada was anxious to explore and the ideas and resources available to her, before we return in the last chapter of this section to Ada herself.

SEVEN

ELECTRIC UNIVERSE

The world of electricity had changed a great deal since the heady radicalism of the years immediately following the Napoleonic Wars. When Tom Weems' remains went under the surgeon's knife and the galvanist's battery in 1819, there was still more than a year to go before the Danish natural philosopher Hans Christian Oersted triumphantly demonstrated the connection between electricity and magnetism, and two years before Michael Faraday at the Royal Institution demonstrated electromagnetic rotation.[1] By 1844, when Ada Lovelace's carriage drew up outside the door at Fyne Court, Faraday's brilliant and meticulous series of researches into electricity and magnetism had been under way for more than a decade. Following his announcement of electromagnetic induction in 1831, Faraday now towered over the world of British science, his reputation overshadowing even that of Sir Humphry Davy, his former master. One of Faraday's great achievements had been to make electricity respectable, divorcing it from its seedy radical origins. For Faraday, this meant excluding electricity from any discussions of the origins of life. There were still plenty of radicals around, nevertheless, and they still had a keen interest in electricity and its relationship with the vital force. With Chartism on the rise throughout the 1840s and revolution simmering not that far beneath the surface of a discontented society, electricity remained dangerous in some quarters at least.

In 1837 Charles Wheatstone and William Fothergill Cooke took out a patent on a new invention – the electromagnetic telegraph – that was set to revolutionise electricity and society.[2] The telegraph heralded a new age of rapid long-distance communication and was the first of the Victorian age's new electrical industries. For many of the new generation of electrical enthusiasts, it was electricity's technological potential, rather than its connections with the mysterious properties of life, that made it a revolutionary science. Through the telegraph, electricity would become 'a dutiful child or obedient servant … traversing the mighty deep in the shape of an angel of peace'.[3] Others were just as rhapsodic about the possibilities of electrical locomotion, imagining electrical trains eating up the miles and electrical ocean liners criss-crossing the Atlantic. Electrical entrepreneurs came up with ideas for electric lights (they were in use in theatres by the end of the 1840s) and electrical detonators for blowing up shipwrecks that endangered shipping. Electricity was touted as a way of growing crops more efficiently. According to one pundit, 'to cross the seas, to traverse the roads, and to work machinery by galvanism, or rather electro-magnetism, will certainly, if executed, be the most noble achievement ever performed by man'.[4] Ironically, given electricity's radical origins, William Fothergill Cooke even suggested that one good use for the electric telegraph would be in the deployment of troops to put down Chartist uprisings.[5]

By the 1840s electricity could mean quite different things to different people. Just what it meant often depended on where it was encountered and who was doing the talking. At the Royal Institution, under Michael Faraday's firm direction, electricity was a carefully sanitised affair. There was not a hint of materialism or political heterodoxy to be found there. Trained originally as a bookbinder's apprentice, Faraday had the good luck to be hired as Sir Humphry Davy's laboratory assistant at the Royal Institution in 1813 and soon joined his new master on a continental tour – Davy was on his way to Paris to receive a medal from Napoleon for services to science. Back at the Royal Institution, by the end of the 1820s he had largely taken over from Davy in the running of the laboratory. Like Davy he too was famous as a lecturer, drawing large crowds to his performances. Writing about life during the 1820s and 1830s many years later, the novelist George Eliot put Faraday nicely in his place in a scathing discussion of fashionable society: 'But then,' she wrote, 'good society has its claret and its velvet carpet, its dinner-engagements six weeks deep, its opera and its fairy ballrooms; rides off its ennui on thoroughbred horses, lounges at the club, has to keep clear of crinoline vortices, gets its science done by Faraday.'[6]

The Royal Institution had gone through some hard financial times during the 1820s, and Faraday's solid management – as well as his rising reputation as a fashionable lecturer – restored its fortunes. He re-adjusted the Institution's lecture schedule so that it fitted in with the fashionable London season. He established the Friday evening discourses to bring in crowds to see bravura demonstrations of the latest discoveries. He also started the annual Christmas lectures. A glance at a famous portrait of Faraday performing at one of these Christmas lectures is enough to establish just how accurate George Eliot's assessment of his appeal was in reality. The lecture theatre was crowded with well-dressed ladies and gentlemen. In the front row, we can clearly see Prince Albert the Prince Consort, with the Prince of Wales on one side and his little brother on the other. Faraday was evidently adept at giving such crowds exactly what they wanted. As one impressionable gentlewoman recalled:

his enthusiasm sometimes carried him to the point of ecstasy when he expiated on the beauties of Nature, and when he lifted the veil from her deep mysteries. His body then took motion from his mind; his hair streamed out from his head; his hands were full of nervous action; his light, lithe body seemed to quiver with its eager life. His audience took fire with him, and every face was flushed.[7]

This was not an experience that many had the opportunity to enjoy. Most people had to find their scientific and electrical thrills elsewhere. There was, however, no great shortage of places where anyone might go to hear about and experience the latest electrical wonders. Right at the other end of the spectrum were radical demagogues preaching sedition on street corners. Electricity and its materialist messages played an important role in their sermonising as well. Even decades later, one of Faraday's biographers remembered coming across an 'infidel preacher' fulminating away on Paddington Green about electricity's subversive potential – and taking the great man's name in vain in the process too.[8] Throughout the 1820s and 1830s radical orators at London's Rotunda and elsewhere regaled their audiences with electrical parables. The fiery publisher and politician Richard Carlile, or when he was in prison his lover and fellow campaigner Eliza Sharples, peppered their political speeches with galvanic language. At radical halls of science all over the country, orators like Thomas Simmons Mackintosh – a follower of the utopian socialist Robert Owen – delivered fiery secular sermons on electricity, life and revolution. Electricity's radical origins might be carefully swept under the carpet in places like the Royal Institution, but they were fully on show elsewhere.

Electricity as it was experienced by visitors to the National Gallery of Practical Science or the Royal Polytechnic Institution was often very different from the variety on show at Michael Faraday's Royal Institution lectures. The Adelaide Gallery (as the National Gallery was popularly known) opened in 1832 on Lowther Arcade and Adelaide Street in London, just off the newly developed Trafalgar Square end of the Strand. It was at the heart of fashionable London. Founded by American inventor Jacob Perkins with a view to popularising his own inventions (one such including a high-pressure steam gun that could fire lead bullets at a target at the rate of seventy balls in four minutes), it soon expanded its remit to include 'whatever may be found to be comparatively superior, or relatively perfect in the arts, sciences or manufactures'. The proprietors promised to display 'specimens and models of inventions and other works &c. for public exhibition'.[9] The Royal Polytechnic Institution that opened its doors a few years later in 1836, in equally fashionable Regent Street, had a very similar rationale. Inventors were invited to put their wares on show and the public encouraged to pay their shillings at the doors to inspect them. Electrical machines and instruments rubbed shoulders with steam-engines, curious mechanical contrivances and strange specimens brought back from the far reaches of the Empire.

Electricity, with its capacity for spectacular display, was an important feature of these exhibition halls' public offerings. A year or so before Ada Lovelace's visit to Crosse at Fyne Court, for example, the Royal Polytechnic Institution shut its doors for a fortnight whilst preparing to unveil its latest electrical acquisition – Armstrong's hydro-electric machine – to its visitors. This gargantuan piece of equipment could be used to mount a truly spectacular extravaganza of sparks and explosions. The press was duly primed to note its substantial dimensions, with a boiler 7.5ft long and 3.5ft in diameter, to produce steam; the steam then came gushing out of a series of nozzles to produce the electricity, which powered the whole contraption. 'The passage of the electricity over the tinfoil on the tubes was far more brilliant, and the aurora borealis exceeded in intensity and beauty anything we had ever witnessed,' the *Morning Chronicle* enthused in September 1843. 'The violet colour was brighter, and at the same time deeper, and the exhausted receiver showed more plainly the progress of the electric spark.'[10] At the Adelaide Gallery, visitors could watch the rare *gymnotus electricus*, or electric eel, being fed twice daily. Both institutions offered displays of new-fangled electric telegraphs, as well as electromagnets, magneto-electric machines and model electromagnetic locomotives.

Closely associated with these exhibition halls of spectacular electrical science was the group of instrument-makers, lecturers and enthusiastic experimenters who made up the core of the short-lived London Electrical Society. Founded in 1837, the London Electrical Society started life as an informal gathering at the instrument-maker Edward Marmaduke Clarke's Laboratory of Science in Lowther Arcade – just across the way from the Adelaide Gallery. Its main mover was the former artilleryman, boot-maker, self-taught instrument-maker and electrical lecturer William Sturgeon, who had made a name for himself just over a decade previously as the inventor of the electromagnet. Members met regularly, initially at Clarke's premises but later in rooms at the Adelaide Gallery, and then at the Royal Polytechnic Institution. A number of the members, including Clarke and Sturgeon himself, had electrical exhibits on show at both venues. Other members of the Society included the consulting chemist William Leithead (who also worked at the London Colosseum's Department of Natural Magic); the railway engineer Charles Vincent Walker; and the wealthy wine merchant John Peter Gassiot. It should go without saying that Andrew Crosse was an enthusiastic member of the Society. It should be equally unsurprising that Michael Faraday was not.[11]

One member of the Society, Henry Minchin Noad, an electrical lecturer based in Bristol and the West Country, waxed lyrical in his *Lectures on Electricity* about the opportunities the London Electrical Society offered to the electrical 'tyro'. The lectures were dedicated to Andrew Crosse: 'to whose indefatigable industry for a long period of years, electrical science is indebted for so rich an accumulation of valuable facts ... and whose liberal, open and communicative spirit is not less remarkable than his enthusiastic love of science.'[12] As well as believing (in principle at least) in scientific collaboration, the Society's members were devotees of display. John Peter Gassiot, the society's treasurer, was well known in London's scientific circles during the first half of the 1840s as the impresario of flamboyant electrical soirées. When the Swiss electrical experimenter Auguste de la Rive visited London in 1843, Gassiot organised a soirée in his honour. Reporting on his experiences in the *Archives de l'Electricité*, de la Rive described how he had been witness to 'some beautiful experiments made by means of voltaic batteries of great power'.

He had seen how 'the voltaic arc, between the charcoal points, lighted up a very large apartment, and made all the wax lights become pale, for their light almost entirely disappeared before that of the electric current'.[13] When the Italian natural philosopher Carlo Matteucci visited London a few years later, Gassiot hosted a similar extravaganza. Charles Vincent Walker in his *Electrical Magazine* marvelled again at the effects produced by Gassiot's huge array of Grove voltaic cells: 'the charcoal light was exceedingly brilliant, – so much so that its reflection was seen at the distance of several miles, and the appearance was believed to be due to some atmospheric meteor.'[14]

All of this goes to show that there were certainly plenty of places (in London at least) during the early Victorian period where people might go to experience electricity. It also suggests that just what they experienced differed quite significantly depending on where they went. The electricity on show at the sober Royal Institution was different from the electricity promoted by radical orators at the London Rotunda, and indeed the electricity that was put on display at the Adelaide Gallery and the Royal Polytechnic Institution was different from that at one of Gassiot's electrical soirées. What people saw – and what they consequently believed electricity to be – varied with place and speaker. Electricity could be dangerously radical or it could be safe and sanitised. It could be the affair of bombastic showmen or the business of respectable lecturers. Relatively few people were in Ada Lovelace's position of being able to move around between these different places and experience electricity from a variety of different perspectives. Ada, on the one hand, was a friend and correspondent of Faraday's. On the other, she associated with radicals. She was familiar with the Adelaide Gallery, having studied the Jacquard Loom on display there for her work with Charles Babbage on his Analytic Engine and, through Crosse, she had connections to the London Electrical Society and its enthusiastic, self-helping membership.

Sitting in the crowded lecture theatre of the Royal Institution listening to Michael Faraday expound his views on electricity, Ada Lovelace would have encountered a very specific and – by the standards of many of his electrical contemporaries – a very idiosyncratic view of electricity and its place in the universe. The philosopher William Whewell remarked in his *Philosophy of the Inductive Sciences* that one of the many virtues of Faraday's ideas was the way in which they were 'unincumbered with extraneous machinery'.[15] When he made the remark in 1840, Whewell was Professor of Moral Philosophy at Cambridge and was soon to be appointed as Master of Trinity College there. In many ways he was the outspoken, intellectual champion of tacit genteel Anglicanism, as was espoused by many of the Royal Institution's audience (and lampooned so mercilessly by George Eliot, as we saw earlier). Throughout the 1830s, following the triumphant discovery of electro-magnetic induction announced in his *First Series of Experimental Researches* in 1831, Faraday had produced series after series of new experimental work. By the 1840s he wanted to argue that his work showed that electricity was lines of force in space.

As Whewell perceptively noted, Faraday's electricity was all about tranquil empty spaces rather than messy machinery.

Like many of his contemporary men of science, including his old master Humphry Davy, Faraday was fascinated by the relationships between the different forces that governed the universe. He was interested in the unity of nature. As William Robert Grove, Faraday's friend and opposite number at the rival London Institution, put it in 1846:

the various affections of matter which constitute the main objects of experimental physics, viz., heat, light, electricity, magnetism, chemical affinity, and motion, are all correlative, or have a reciprocal dependence ... neither, taken abstractedly, can be said to be the essential or the proximate cause of the others, but that either may, as a force, produce the others.[16]

Noticeably absent from Grove's list of forces – and it was most certainly absent from Faraday's list too – was the life force. Faraday, Grove and others like them from a new generation of electrical experimenters and theorisers were trying to make electricity, and the language of unified nature that went along with it, respectable. That meant that they had to strip it of its radical connotations and divorce it from its original connections with materialist ideas about the origins of life. The spectre of the French Revolution still loomed large over British electricity, even as late as the 1840s, and even the slightest hint of materialism or radicalism was fiercely resisted by the new breed of gentlemen of science.

This is very clear, for example, in Faraday's own single explicit foray into the debate about electricity and life, in the aftermath of Andrew Crosse's notorious experiments in 1837. For his *Fifteenth Series of Experimental Researches*, Faraday carried out a number of experiments using the Adelaide Gallery's famous electrical eels. The fact that Faraday was prepared to take the highly unusual (for him) and risky step of dabbling with the populist Adelaide Gallery at all, hints at his sense of the experiments' urgency. He had no choice though – the Adelaide's eels were the only living ones in London. In the first instance, the experiments were meant to demonstrate that the electricity produced by the eels was the same as electricity produced from other sources, such as from galvanic batteries or electrical machines. In this respect the experiments were simply an extension of work that Faraday had already done in demonstrating the identity of electricity from different sources. But Faraday wanted to do more than that. He wanted to use his experiments with the electric eels to draw a line in the sand. His *Fifteenth Series of Experimental Researches* quite explicitly set out to define the limits of any legitimate inquiry into the origins of life.

His experiments, he said, were 'upon the threshold of what we may, without presumption, believe man is permitted to know of this matter'.[17] Whilst they did indeed show that 'the nervous power is in some degree analogous to such powers as heat, electricity, and magnetism', all this meant was that the nervous power itself was not and could not be 'the direct principle of *life*'.[18] This was drawing an interesting and fine distinction, and it is worth noting that this was a distinction in principle rather than one based on any kind of observation. As far as Faraday was concerned, if the nervous power could be experimented with, then this meant that it was not, ultimately, the force of life itself. Life for Faraday was, by definition, beyond any kind of mere human meddling. That meant that anything that turned out as a matter of fact, amenable to experimental investigation, simply could not be the principle of life. This was not just moving the goalposts, it was pulling them down and taking them off the playing field completely. The simple fact that electricity could be experimented upon meant for Faraday that it could not be the principle of life. And if the nervous power was amenable to experiment as well, then it was not the ultimate principle of life either.

If Faraday's electric universe was 'unincumbered with extraneous machinery', the same can hardly be said about the electrical worldview espoused by William Sturgeon and many of his fellow experimenters at the London Electrical Society. Far from it; Sturgeon's universe was full of electrical machinery. Sturgeon excelled at producing electrical apparatus that mimicked nature. In one of his earlier experiments, for example, a globe made out of silver and bismuth wires was made to rotate around a central magnet when the junctions of the different wires (located, appropriately enough, at the artificial globe's equator) were heated, thus producing electricity. Sturgeon speculated that this was the way the Earth's rotation took place as well. The experiment, he claimed, was 'obviously analogous to the natural state of the earth' and explained the prevalence of thunderstorms in the tropics as well as the Earth's rotation. It showed how 'the action of the sun either partly or wholly governs the general electrical phaenomena of nature; and, either by producing or exciting this wonderful agency in the equatorial regions, dispenses its influence from thence to the poles of the earth'.[19] The universe was 'Nature's laboratory', and it was 'well stored with apparatus of this kind, aptly fitted for incessant action, and the production of immense electrical tides; and the insignificancy of our puny contrivances to mimic nature's operations, must be aptly apparent when compared with the magnificent apparatus of the earth'.[20]

Sturgeon's fellow members of the London Electrical Society certainly managed to produce plenty of experimental evidence that appeared to back this view of a universe teeming with electricity. At almost every meeting of the London Electrical Society during its short-lived existence, the Sandwich-based surgeon and electrician William Henry Weekes announced the results of his monthly experimental surveys of the state of atmospheric electricity. His apparatus consisted of wires strung from nearby rooftops and church spires. Electricity collected from the atmosphere in this way could be used to devastating effect. Whenever storm clouds gathered, 'pregnant with infuriated lightnings, and momentarily gaining additional sublimity from reverberating peals of deafening thunder', Weekes was witness to:

tremendous torrents of electric matter, assuming the form of dense sparks, and possessing the most astonishing intensity, rush from the terminus of the instrument
with loud cracking reports, resembling in general effect the well-known running
fire occasioned by the vehement discharge of a multiplicity of small firearms. Fluids
are rapidly decomposed; metals are brilliantly deflagrated; and large amounts of
coated surface repeatedly charged and discharged in a few seconds.[21]

Andrew Crosse conducted similar experiments at Fyne Court, to the dismay of his
neighbours. Experiments like these seemed to provide irrefutable evidence that
electricity went right to the heart of nature's operations.

If electricity was everywhere above the Earth's surface, it was everywhere under-
neath it as well. The experiments for which Crosse was feted onstage at the Bristol
meeting of the BAAS in 1836 – and which ultimately produced his equally infamous insects – had been designed to show that crystals were formed underground
by electricity. The Cornish Quaker Robert Ware Fox was similarly convinced of
the role of electricity in forming veins of minerals in the earth. Fox, like Sturgeon,
argued that atmospheric and subterranean electricity alike were part of the same
great pattern. The aurora, for example, was 'an exhibition of electric currents at
a great height, which are connected with others nearly parallel to them, in the
interior of the earth'.[22] Fox thought all this was evidence of providential design;
electricity made sure useful metals were helpfully concentrated in places where
they could be mined economically, rather than being diffused throughout the
Earth. Another London Electrical Society member, Thomas Pine, experimented
with using electricity to improve crop yields. He too thought that electricity was
everywhere, 'whether accumulated in the natural artillery of the heavens – in the
volcanic bowels of the earth – or concentrated in the artificial arrangements of
a voltaic battery'.[23] Electricity was the key both to understanding nature and to
exploiting it.

Accounts like these of the all-pervasive nature of electricity were grist of the mill
for radicals who wanted to underpin their politics with an impeccably materialist
philosophy based on the latest science. The flamboyant inventor Francis Maceroni
– whose colourful past included acting as aide-de-camp to Napoleon's henchman
and King of Naples, Joachim Murat – was one who saw electricity everywhere.
During his years in Naples as Murat's sidekick, Maceroni had dabbled enthusiastically in medicine, mechanics and natural philosophy. In England during the
1820s and 1830s, following an unsuccessful period as a mercenary soldier in South
America, Maceroni gained some initial success as an inventor, although the railways killed off his efforts to market his patented steam coach. Notoriously, in the
febrile atmosphere of the months leading up to the passing of the Reform Bill in
the summer of 1832, Maceroni published a pamphlet describing how to construct
homemade weapons for street fighting.[24] Recalling, at around this time, his exploits
in Italy in the pages of the *Mechanics' Magazine*, Maceroni speculated about the links

between electricity, volcanic eruptions and the state of the atmosphere. He thought that layers of air, clouds, water and earth were like the alternating layers of a galvanic battery – or like the body of an 'organized being'. The 'brains, nervous ganglions, and nerves, which are evidently the seat of vital action, in the identities we call animal', were 'real electrical machines'.[25]

This kind of enthusiastic appropriation of electrical experiments and ideas was far from unusual in radical circles in the years surrounding the passing of the Reform Act in 1832. The Lady of the Rotunda, Eliza Sharples, sermonised to enthusiastic crowds on 'how far the Human Character is formed by Education or External Circumstances'. Informing her listeners that the 'cause of the *sentient* principle in animals being more peculiarly in the brain is, that it is an electric pile giving pulsations to the heart and accounting for all the phenomena of the body'. The best definition of 'the human body and its life, and of animal life generally' was that it was 'a self-acting electrical machine, sustained by currents of atmospheric air and liquids'.[26] Discoursing on the Bible a few months later, she told her audience that the:

> conceit, that spirit can retain an identity without the aid of the body, is that of superstition and madness … the best account that has yet been given of life is, that it is electric action. All electricity depends upon certain arrangements of materials, without which it cannot exist; so that the imagination of life without body, is like the creation of all things out of nothing.[27]

In other words, just as electricity could not exist without some machine to produce it, so life could not exist without a body to sustain it.

Thomas Simmons Mackintosh developed his *Electrical Theory of the Universe* around similar themes. Mackintosh's universe was full of electricity. The sun was a huge conductor charged with electricity, whose influence pervaded the entire solar system and dictated the movements of the planets. He turned to Crosse, Fox, Pine and Weekes to provide him with evidence of electricity's pervasive influence in terrestrial matters and to show that 'electrical action is incessantly going on in the animal, vegetable, and mineral kingdoms, that animals, vegetables and minerals are produced and maintained each in its own appropriate condition' by means of electricity.[28] What all this meant was that is was 'better to view man as an organised machine, and to search for the seat of those impulses in the functions of his physical nature, where assuredly they are to be found, than to trace them to sources beyond our knowledge and above our control'.[29] Electricity demonstrated mortality and the need for bread now rather than cake tomorrow. In just the same way that clocks ran down or galvanic batteries ran out of their electric fluid, 'life is a process which only exists by a continual approach towards death'. The electrical universe meant that 'eternal life and perpetual motion are almost, or altogether, synonymous' – that is to say, impossible.[30] Time therefore, for revolution on earth since there was no hope of heaven. Mackintosh was a committed follower of the utopian socialist Robert Owen, founder of New Lanark in Scotland and New Harmony in the United States as socialist co-operative experiments.

Not all producers of universal theories of electricity were diehard radicals like Maceroni, Sharples or Mackintosh. The notorious weathermonger Patrick Murphy maintained that electricity was 'the *vivifying principle* which imparts and diffuses over the otherwise inanimate masses of the sun and planets, the spirit of vitality which, as we perceive, pervades and quickens their elements', in order to 'develop and sustain the germs of the animal and vegetable kingdoms, of which, similar to the earth, the entire of these bodies are assumed to be the theatres'.[31] So electricity was the force that spread vitality through the universe – and made extraterrestrial life a certainty. William Leithead, member of the London Electrical Society and the London Colosseum's superintendent of the Department of Natural Magic, held that electricity was 'incomparably the most active and momentous principle in the whole physical world'. It possessed 'a celerity of motion beyond that of light' and 'a chemical power superior to caloric'.[32] Electricity was everywhere. It caused earthquakes. It caused cholera. No motion could take place in the universe 'without being accompanied by a disturbance in the equilibrium of the electric fluid'.[33] Leithead maintained that 'it is probable that gravitation, cohesion, and the other species of attraction, are all modifications of the attractive power of electricity'.[34]

Far more dangerous, as far as many of the gentlemen of science were concerned, was the role electricity played in the bestselling and controversial *Vestiges of the Natural History of Creation*, published anonymously in 1844. *Vestiges* caused a real stir amongst the scientific elite, with its description of a universe developing according to the dictates of natural law. Adam Sedgwick, Cambridge's Woodwardian Professor of Geology, thought the book's arguments so outrageous it must have been written by a woman. The woman he had in mind was, of course, Ada Lovelace. *Vestiges* was in reality written by a man, an Edinburgh publisher, Robert Chambers. Chambers made use of the latest scientific researches to underpin his theory and turned to electricity for much of his evidence. 'Electricity we also see to be universal,' declared Chambers, and so if 'it be a principle concerned in life and in mental action, as science strongly suggests, life and mental action must everywhere be of one general character'.[35] Electricity, said Chambers, was 'almost as metaphysical as ever mind was supposed to be', but was at the same time 'a real thing, an actual existence in nature'. All of which tended to show the 'absolute identity of the brain with a galvanic battery'.[36]

Chambers' account of a universe governed by a progressive natural law was meant to underpin liberal, reformist politics rather than the fiery radicalism espoused by Sharples or Mackintosh.[37] The system of electrobiology developed during the 1840s by Alfred Smee was similarly meant as an antidote to revolutionary sentiment and a way of defending orthodox, Anglican religion. Smee, like Chambers, envisaged the brain as a galvanic battery, linked to the peripheries of the body by 'bio-telegraphs.'[38]

Everything about the body worked by electricity, and Smee went to great lengths to suggest how different electrical arrangements might be used to model the senses of hearing, sight, smell, taste and touch. The idea of God, according to Smee (and he had in mind the orthodox and conservative Anglican God) was hard-wired into the electrical structure of the brain. Catholics, Chartists and other ne'er-do-wells were quite literally suffered from faulty wiring, resulting in 'an irregular action of the brain, analagous to the irregular action of machinery'.[39] Unfortunately for Smee, the link between electricity and radical politics was too firm in most readers' minds for them to readily accept his conservative version of galvanism. Reviewers simply poked fun at his efforts to build a 'living, moving, feeling, thinking, moral, and religious man, by a combination of voltaic circuits'.[40]

All of this shows that someone like Ada Lovelace, anxious to use electricity to make sense of their own body, could look in many different directions for their inspiration. At one extreme, she could follow Michael Faraday's lead – along with most of her gentlemanly scientific acquaintances – and draw a firm line between investigations of the body's electrical properties and the experimental investigation of life itself. At the other end of the spectrum of belief were the grandiose claims of out-and-out materialists and radicals like Sharples and Mackintosh, who had no time for any such distinction and took electricity to be the material stuff of life and the driving force behind the universe. Radicals like these could point to Sturgeon's electromagnetic reproductions of natural phenomena, Weekes' and Crosse's experiments with atmospheric electricity, Thomas Pine's efforts at electro-cultivation and – of course – the *acarus crossii*, as evidence to underpin even their most apparently outlandish claims. When Lovelace set out for Fyne Court in November 1844, Chambers' *Vestiges of the Natural History of Creation* had been in the booksellers' shops for only a few weeks. Ada may well have already read it (and might even have been aware of the rumours circulating that she was its author). She would certainly have known that in seeking to turn her body into an electrical laboratory, she was dipping her toe into some very controversial waters indeed.

EIGHT

GALVANIC MEDICINE

In the preface to his *Elementary Lectures on Galvanism*, the electrician William Sturgeon described his own attempts at raising the dead with electricity. His opportunity had come when he was brought 'the bodies of four young men, who were drowned by falling through the ice, at Woolwich'. They had clearly been handed over as a last resort – Sturgeon complained that two hours had been wasted on 'the usual routine of medical treatment' before he had the opportunity to try his batteries. By this stage there was little that he could hope to achieve, but Sturgeon noted nevertheless that three of the corpses were 'still alive to the Galvanic influence'. Some of them 'opened their eyes and moved the lips; and one of them bent the elbow when the current traversed the arm from the shoulder to the hand'.[1] Sturgeon was certainly not the only electrician to try his hand at electrical medicine. Andrew Crosse made therapeutic use of his electrical apparatus too. By the beginning of Victoria's reign, medical electricity had a history that stretched back for nearly a century. It made sense, after all, to think that if electricity and the nervous force were one and the same, electricity might be used to replenish exhausted bodily reserves. But this was exactly where the problem lay for early Victorian electrotherapists, because it seemed to depend on the assumption that electricity was deeply implicated in the processes of life, a claim now most commonly associated with materialists and radicals; their practice looked like materialism too.

So despite its long history, medical electricity at the beginning of Victoria's reign was increasingly a marginal practice. Its practitioners were easily tarred with the brush of political heterodoxy, or dismissed as quacks and charlatans, as Doctor Caustick had tarred Aldini's medical assistants forty years earlier. Electricity was used as therapy in some hospitals where an individual doctor happened to be an enthusiast, but by and large, medical electricity existed outside the medical mainstream. It was the province of electricians rather than medical men. There continued to be a thriving trade in galvanic belts and other kinds of electrical medical apparatus, but that hardly did much to raise the status of medical electricity for a medical profession that was busily trying to sever any remaining links between medicine and mere trade. Medical electricity looked more like flashy showmanship, shady dealing or rank political heresy than rational therapeutics to many doctors' eyes.

However there were some medical practitioners during the 1830s and 1840s that were determined to go against prevailing respectable opinion within the profession. They would try to reclaim electricity for medical orthodoxy and wrest it from the hands of electricians unqualified to meddle with the workings of the body, or charlatans willing to do anything that would show a profit. To do this, they needed to find ways of redefining the relationship between electricity and the body, and create a new kind of space for galvanic medicine.

The controversial physician Marshall Hall argued from the 1830s onwards that his experiments demonstrated the existence of an 'excito-motory system' directed from the spinal cord. This 'true spinal system' was the 'seat or nervous agent of the appetites and passions, but is also susceptible of modification by volition'.[2] According to Hall many, if not most, of the nervous system's operations were automatic, governed by his 'true spinal system' rather than the brain. This separate automatic system governed much of the body through reflex action. One implication of Hall's theories was that the nervous system could be subjected to experiment without raising difficult questions about the relationship between electricity, the nervous force and the will or the soul. Yet even this was a step too far for the old guard at the Royal Society, particularly since Hall's research also suggested that the role of the conscious will in controlling the body was seriously curtailed. After publishing the first of Hall's experimental accounts in 1833, the *Philosophical Transactions* refused to publish any more. The backlash was led by Peter Mark Roget, one of the Society's two secretaries. It was not until a new generation of reformers captured the Royal Society in a virtual *coup d'état* in 1848, and sent Roget packing in the aftermath, that Hall had any luck in persuading the *Philosophical Transactions* to accept his contributions.[3]

Hall was enough of a self-publicist to make the most of the Royal Society's cavalier treatment of his researches, however, and he had his own allies. There were other scientific journals such as the *Philosophical Magazine* or the *Edinburgh New Philosophical Journal*, as well as medical journals like the *Lancet* or the *Medical Gazette*, which might be less prestigious than the *Philosophical Transactions*, but were just as effective in getting his message across. One ally, the Welshman Thomas Williams, had been a tutor at Guy's Hospital before joining the radical surgeon Richard Grainger's Anatomy School on Webb Street. He helped Grainger perform the public dissection of Richard Carlile's body in 1843.[4] Williams followed Hall's researches on the excito-motory system in his own experiments, paying tribute to 'the labours of a man who, for the last thirty years, has laboured with unquenchable zeal to destroy the trammels of error and mystery in which practical medicine was involved'.[5] Translated in the *Philosophical Magazine*, Charles Vincent Walker's *Electrical Magazine* and other publications, the researches of continental investigators like the Italian Carlo Matteucci in electrophysiology were also gaining a readership. Matteucci used his galvanic apparatus to probe the electrical properties of nerves in detail whilst firmly denying the identity of electrical and nervous force. Matteucci visited England in 1845 and performed convincing demonstrations of his experiments in front of audiences containing the likes of Michael Faraday and Marshall Hall. Experiments like these seemed to provide a framework for making sense of electricity's therapeutic possibilities whilst avoiding the materialist pitfalls.[6]

At Guy's Hospital in London, determined efforts to make electrotherapeutics respectable had been underway since the middle of the 1830s. In 1836 the hospital established an 'electrifying room' to cater for patients who were to receive electrical treatment. In charge was an ambitious young physician, Golding Bird, anxious to make his mark on the hospital and the wider medical and scientific worlds. Bird was in a good position to make such an impact; he had attracted the patronage of some of the hospital's senior medical men – the surgeon Astley Cooper and the physician Thomas Addison. Thanks to their patronage, he had been appointed first as a lecturer in natural philosophy to Guy's Hospital medical school, and later as assistant physician. He was an avid chemist and a member of the London Electrical Society. He had just the right kinds of skills and contacts to make the electrifying room a success. His ambition at Guy's was to turn electricity into hospital medicine, administered by trained and experienced hands for the treatment of specific diseases. He complained that in the past electricity had appeared to some 'to possess an almost magical action in the most intractable of diseases; whilst others, equally worthy of confidence, have declared it to be utterly useless'. Too often, he said, the cry 'Let them be electrified' had been uttered more as a last resort than 'from any well-defined view of its real influence'. The Guy's experiment was going to change all that, he hoped.[7]

Electricity was administered in a number of different ways in the Guy's Hospital electrifying room. One common method was to sit the patient on a stool with legs made of glass, or some other non-conducting material, and connect them to the prime conductor of an electrical machine. The result was that the patient's 'whole surface becomes, by induction, powerfully electro-positive; and a combination, by silent discharge, is constantly occurring, by absorption of negative electricity from the air: this may be seen to take place luminously in the dark, especially about the hair, eyelashes, fingers, &c.'.[8] This was called an electric bath and was thought to work by heating the body and improving secretion. A variation of this treatment was to hold an earthed conductor close to the patient's body whilst they were undergoing an electric bath, 'producing a vivid flash of light and a snapping noise'. Electric shocks were also administered, either from Leyden jars, from galvanic batteries or from magneto-electric machines. Bird's favourite device consisted of a:

> bundle of soft iron-wire … Placed in the centre of a wooden reel or bobbin; around which is wound a quantity of stout insulated copper-wire, having its two ends free, for the purpose of being connected with a single pair of plates of copper and zinc exposed to chemical action: over this is wound about 1300 feet of very thin insulated copper-wire.[9]

Induction coils like this one were to be commonly used for medical purposes, both inside and outside hospitals, for the rest of the century; a use that could scarcely have pleased Michael Faraday, the discoverer of electromagnetic induction.

From the late 1830s onwards and for the next fifteen years or so, *Guy's Hospital Reports* published regular accounts of the electrifying room's activities and of the patients treated within its walls. From the beginning, the most common disease treated was chorea, popularly called St Vitus' dance. In his first report Bird listed thirty-six cases, apparently caused by anything from 'terror from confinement in a cellar' to menstrual abnormalities to grief. The usual treatment was to draw sparks from the spine and to administer electric shocks.[10] The patients varied in age from 8 to 61 years old. The majority, however, were young women between the ages of 12 and 20. The same general pattern of patients and treatments can be seen in later accounts from the Guy's electrifying room up until the early 1850s.[11] William Gull in 1852 was simply repeating the orthodox view developed at Guy's when he commented that 'electricity acts here as any other mechanical stimulus', and that it worked primarily by simply 'stimulating the functions of the nerves and muscles'.[12] The patients were all from the respectable working class, of course. No self-respecting member of the middle or upper classes would ever be found as a hospital patient. Reputations forged on the hospital wards could nevertheless prove the foundation of an extremely lucrative private practice.

Doctors like Golding Bird insisted that electrotherapy needed medical expertise. Medical men, not electricians, should be the ones trusted to administer electricity to the body. Indeed this was to be a constant refrain for hospital-trained electrical medicine enthusiasts for the rest of the century. Electricity was a tool for the trained doctor, not some jumped-up electrical specialist, said Bird, arguing that 'there is no more necessity for increasing the "specialities" of our profession by the electrician or galvanist, than by a qualified administerer of clysters'. According to him, 'I have never needed any other electrician than the wife or mother – often the footman or maid-servant'.[13] Bird had his beady eye on the unorthodox competition, whose activities swamped anything that he or his colleagues at Guy's could offer. Every guinea paid to these men was 'a direct robbery committed upon the regular and educated practitioner'. Bird complained that medical galvanists who 'blazon their electrical and galvanic nostrums as a panacea for every ill' were fleecing the public and depriving doctors of their fair share. He knew of one who 'pockets some thousands per year by his practice'. Their numbers were on the increase, he noted, suggesting that 'a new class of "medical electricians" have of late started up, in the shape of the persons employed at some of the railways to exhibit the electric telegraph'.[14]

William Hooper Halse would certainly have been one of the subjects of Golding Bird's wrath, though given his links with William Sturgeon and the London Electrical Society – of which Bird was certainly a member – it is not impossible that the two men knew each other. Not only was he a self-styled medical galvanist raking in his thousands at the expense of hard-working medical men like Bird, but he exhibited just those dubious links with shady radical characters that had given electricity and electricians such an unsavoury reputation in some quarters. Halse had arrived in London around 1840 from Devonshire at the behest of the infidel publisher Richard Carlile and his mistress Eliza Sharples. Carlile was anxious to have Halse settled in the metropolis precisely because of his reputation as a medical galvanist. Halse had links with the London Electrical Society; he published a number of accounts of his electrical experiments in William Sturgeon's *Annals of Electricity*, including one in which he described his attempts to revive a number of drowned puppies by electricity. Halse took his experiments to be additional ammunition for those who wanted to exploit the 'astonishing powers of galvanism on the human frame, in supplying the nervous fluid (or a substitute for it) and the ignorance of this fact by a large proportion of the medical profession'. And he humbly took advantage of the opportunity to bring his own 'quite inexpensive' galvanic apparatus to the public's attention.[15] Halse dabbled in mesmerism as well – another practice in which Carlile and Sharples were interested. His posters, announcing him as 'Proprietor of the Galvanic Family Pill' promised that his lectures at the City Hall on Chancery Lane would reveal God as 'the Triune Magnet of Universal Attraction, generative of Peace and Love'.[16]

Halse's pamphlets were full of recommendations for his galvanic treatments from all kinds of sources. He boasted that he received his patients through 'the recommendation of the principal physicians of the metropolis' and was 'now patronised by a large number of the aristocracy'. His therapy was effective in curing:

> all kinds of nervous disorders, asthma, rheumatism, sciatica, tic doloreux, paralysis, spinal complaints, long-standing headaches, deficiency of nervous energy, deafness, dulness of sight, liver complaints, general debility. Indigestion, stiff joints, epilepsy, and recent cases of consumption.

The reason it worked was quite simple. It was clear that 'the galvanic apparatus can not only supply every purpose of the nervous influence, but that the fluid produced by the apparatus and the nervous fluid are one and the same thing'.[17] Halse offered his services at a guinea for half an hour. He also offered his portable galvanic apparatus for sale at 10 guineas. His apparatus was 'the height of perfection', and using it like:

> some mighty magician of old, Mr. Halse regulates the stupendous batteries around him with the utmost ease and simplicity, directing their curative and health-giving, yet invisible streams, to the afflicted or paralysed joint or nerve, as the case may be, and restoring the one or other to motility or sensibility.[18]

According to the Liverpool medical galvanist Dawson Bellhouse, it was 'acknowledged by medical men of eminence, both in this country and abroad, that there is no remedy more powerful, and beneficial in its effects, than Medical Galvanism, when properly administered'. Like Halse, he offered reasonable terms – in this case 12s 6d (payable in advance) for six applications. His wife was available to provide the treatment for ladies at their own residences, if necessary. Different versions of his apparatus were on sale at prices ranging from £10 10s to £1 10s, though even Bellhouse acknowledged that he could 'not recommend' the cheapest apparatus for 'medical use'.[19] Most such galvanists offered to cure a similar range of ailments with similar therapies, though some, like a Mr W. Hardy of Harrogate seem to have specialised in specific treatments – in this case electro-chemical baths suitable, presumably, for the clientele at a fashionable spa town. They also tended to share a scorn for the orthodox profession, though were perfectly willing to accept any testimonials (inadvertent or otherwise) that came their way from that quarter. If Bird was ready to damn Halse's lack of skill, Halse was just as willing to reciprocate. As far as he was concerned, most members of the medical profession (not to mention his galvanic competitors) knew 'no more how to manage it than a donkey knows how to manage a musket'.[20]

The marketplace was increasingly flooded with galvanic remedies as competing practitioners battled for attention. In 1845 Charles Vincent Walker's *Electrical Magazine* drew sardonic attention to the latest fad for Dr Graham's galvanic rings made of copper and zinc. The usual list of nervous ailments could be cured simply by wearing the ring 'so as to fit close, but not too tight'. Poor Carlo Matteucci should simply give up, go home and 'be silent for ever', Walker felt. Tongue still firmly in cheek, he gave his view that a ring of 'plain gold, placed on the fourth finger of the left hand' was the one that usually had most effect on the nervous system.[21] He might have been aware that there was even a popular ballad to that effect:

> We've Galvanic *Rings*, too, – which, no doubt are fine things, too –
> But *our fair* heed not aught half so worthless and cold;
> *They'd* much rather you sing now, – 'Let's play "Kiss in the Ring" now'
> And be teased to wear for you the *ring* – of 'PLAIN GOLD!!'[22]

Galvanic rings and galvanic belts were becoming ingrained in the new Victorian age's growing commodity culture. They complemented the 'marvellous machinery of man himself' and kept everything ticking over.[23]

Trying to make sense of this marvellous machinery, J.O.N. Rutter turned to the latest electrical technology. 'Electricity being so closely associated with vitality,' he pondered, 'is it not probable that it exercises a powerful influence over the operations of the mind? May not the brain be like a central telegraph-station – a medium of communication between the mental and the corporeal?'[24] Rutter had developed an instrument – the galvanoscope – that could measure this elusive human electricity. This was a simple enough device, consisting of a sensitive galvanometer (an instrument for measuring the strength of electric currents) connected by two lengths of wire to electrodes immersed in bowls of water. The frontispiece to Rutter's book shows a demure young woman demonstrating the apparatus. When both hands were placed in the bowls of water and the right hand clenched, the galvanometer indicator would move in one direction. When the left hand was clenched, it would move in the other. Good health meant being able to generate a healthy current. It meant that all the organs were in 'a state of equilibrium; their powers being so well-adjusted and accurately balanced, as to produce the greatest quantity of vital energy, with the least amount of wear and tear of material'.[25] Sensitive nerves made good conductors, Rutter thought.

Because the human body was such a sensitive indicator for the presence of electricity in the surrounding atmosphere, Rutter was sure that electricity was the root cause of many diseases and epidemics. He believed that in 'an impure atmosphere … the demand upon the electrical mechanism (?) of the body may be so great, and so suddenly made, as to produce such a degree of exhaustion of the vital energies, as will result in disease and death'.[26] He was thinking about the recent outbreaks of cholera, and he was not the first to make the link between cholera and electricity, either. William Leithead had taken a similar view fifteen years or so earlier. He had suggested that epidemics could not be caused, as prevailing medical theories suggested, by atmospheric miasmas, because if this were the case they would attack everyone at the same time; poisons affected everybody regardless of the strength of their respective constitutions. Epidemics like cholera were instead thought to be caused by 'the peculiar electric state of the atmosphere' which did affect individuals differently depending on their sensitivity to the electrical influence.[27] When asiatic cholera had struck the country, 'thunderstorms were more than usually frequent and violent' and 'a very short time before the irruption of the disease, it was observed that the nights were characterised by a highly electrical state of the atmosphere'.[28]

Epidemics and earthquakes happened together, according to Leithead, both produced by what he called the 'pestilential principle' of electricity. Reviewing Leithead's book in the *Philosophical Magazine* (and, incidentally, accusing him of plagiarising the natural philosopher Peter Mark Roget's writings), one anonymous author thought this was all 'neither more nor less than an electrical dream of Mr. Leithead's fancy'.[29] To Leithead, on the other hand, it made perfect sense. Epidemics were caused by 'a diminution in the natural quantity of electric fluid generated in the system, during the more feeble chemical action – exposure to a negative atmosphere – and an impaired conducting power in the nerves'.[30] To Sir James Murray, a decade or so later, the electrical nature of epidemics provided a clue to their cure. Different kinds of epidemics were caused by either an excess or a deficiency of electricity in the atmosphere. Murray suggested that this could be alleviated by providing electrical insulation for housing, as well as lightning rods to carry away the excess of galvanic fluid.[31] All these theories operated on the common assumption that humans were electrical creatures, moving around in an environment entirely pervaded by electricity. Everyone was swimming in a sea of electric fluid. Schemes for insulating houses against the cholera, as much as offers of electrical therapies, belts and galvanic rings, were presented as ways of managing the relationship between bodily electricity and the electricity filling the surrounding universe.

Some kinds of bodies, however, were universally acknowledged to be more susceptible to the vagaries of electricity than others. The female body was particularly sensitive to electricity and at risk, therefore, of the consequences of the electric fluid being out of place. It was commonplace throughout the first half of the nineteenth century to regard women's bodies as being, in some sense or other, out of control. Men's bodies on the other hand were, or were meant to be, firmly under the control of the mind. That was one reason why mental discipline mattered so much to the early Victorians – and why gentlemen of science like Babbage or John Herschel thought that self-discipline was a pre-requisite for science as well. The mental control and discipline they thought was required to be a man of science was the epitome of masculinity. In a woman's case, commentators argued, it was more often the other way round. This was evidently why women could not be practitioners of science, too; they were at the mercy of their bodies. The idea that hysteria was both a typically female dysfunction and was caused by dysfunctional sexual organs was an old one (indeed the word hysteria is derived from the classical Greek word for uterus). The Victorians, however, gave it a new gloss by making it an important part of new theories about the functioning of the nervous system and about the role of electricity in the body.

According to J.G. Milligan, the 'tyrranical influence' of 'corporeal agency' would 'frequently cause the misery of the gentler sex'. Any woman was 'more forcibly under the control of matter; her sensations are more vivid and acute, her sympathies more irresistible'. Women were 'less under the influence of the brain than the uterine system, the plexi of abdominal nerves, and iritation of the spinal cord'. As a result, 'a hysteric predisposition is incessantly predominating from the dawn of puberty'.[32] The Paris-trained physician James Henry Bennet was one of a number of medical men forging a speciality in the diseases of women during this period, and he took much the same view. According to him, even the:

> most intellectual and strong-minded women are not exempt from this reaction of the uterine disease on the nervous system. Under its influence they become irritable and capricious, without the slightest suspicion being entertained by those around them as to the cause of the change that has taken place in their mental state.[33]

Views like these depended, to some degree, on new theories of automatic nervous action of the kind promoted by Marshall Hall. Hall himself was sure as well that hysteria was an 'affection of the excito-motory system', and that it depended on 'the state of the intestine or uterus'.[34] He put the difference between men and women succinctly: 'In the female sex, it is at the moment when disappointments in love are most apt to take place; and in the male, when disappointments in projects of ambition are most apt to occur, that we most frequently observe insanity.[35]

Edward John Tilt – another Paris-trained physician hoping to specialise in female diseases – suggested that 'the principal reason why the knowledge of diseases of women is perhaps so little advanced, is the fact that one sex only is qualified by education and powers of mind to investigate what the other sex has alone to suffer'.[36] He had no doubt that the 'influence of the ovario-uterine organs on the brain and on the mind is unanimously admitted; likewise that this influence is often morbid'.[37] The uterus acted on the brain through the medium of the 'ganglionic nerves' – which Tilt argued played a more important role in female than male physiology. As a result, whilst men had more animal power:

> there is a greater amount of vegetative power in women, for while the proper development of the testicles at once immutably imparts its characteristic effects to man, – the noblest of created beings, – in women, the corresponding organs react more strongly on her system during the reproductive period of life, subjecting it to constant vicissitudes of health and disease.[38]

Another doctor made the electrical mechanism underpinning such views quite explicit: 'The ovaria, uterus and mammae [breasts] form, as it were, a reproductive pile, the circuit being completed by the nervous system'; women's bodies were just like galvanic batteries.[39]

Not all budding specialists in women's diseases shared the view that hysteria had physical origins. As far as Robert Brudenell Carter was concerned, hysteria was entirely a dysfunction of the emotions and had no clear physical basis. It usually had its origins in suppressed sexual desire, or 'any circumstances which direct attention to the reproductive system'.[40] According to Carter, hysteria had three stages. The first was caused by some strong internal emotion – usually associated with sex. Secondary hysteria occurred as a result of a reflex action – remembering or repeating the original emotions would bring about a repetition of the original hysterical attack. Tertiary hysteria, on the other hand, was self-induced – it was simply a deliberate attempt to gain attention and sexual gratification. Girls, according to Carter, had learnt that hysteria was a good attention-seeking strategy and brought its own rewards. 'It is scarcely possible at present,' he complained, 'for an hysterical girl to have no acquaintances among the many women who are subjected to the speculum and caustic, and who love to discuss their symptoms and to narrate the sensations which attend upon treatment.'[41] It was a moral disorder and required strictly moral treatment. In fact, the last thing a doctor should do was pay any attention at all to the patient's genitalia since that would simply pander to her depraved appetites.

Robert Brudenell Carter's views on the moral context of hysteria were in many respects similar to those of Thomas Laycock. Laycock was probably in the vanguard of those Parisian-trained practitioners who tried to make names for themselves as experts on female diseases during the 1830s and 1840s. As far as he was concerned, Laycock suggested, hysteria was to be found affecting 'in some one of its varied forms, almost every female'.[42] Unlike Carter, he had no doubt that it had a straight-forward physical origin in the reproductive organs. It had physical symptoms as well. Laycock suggested that 'very plump mammae, prominent nipples, and dark-tinted areolae are so frequently present in young hysterical females, as to aid with other symptoms in forming a diagnosis'.[43] Along with this 'hysterical embonpoint' came further physical symptoms:

> the appetite is much impaired or abolished, the most minute portion of food, and that only farinaceous, being taken for months together. Yet, to the great surprise of every one, the limbs, mammae, and trunk, continue round and plump. Indeed the most common subjects of hysteria are those endowed with this brilliant plumpness of the surface and delicacy of finish.[44]

Laycock even went so far as to suggest that hysterical women had the general appearance of Parisian prostitutes – something with which he was presumably inti-mately familiar since his time there as a medical student during the early 1830s.

Laycock was sure that even though the immediate cause of hysteria could be located in a derangement of the sexual organs affecting the brain by reflex through the nervous system, none of this took place in a moral vacuum. Hysteria was the sign of an excessively permissive society: 'Young females of the same age, and influ-enced by the same novel feelings towards the opposite sex, cannot associate together in public schools without serious risk of exciting the passions, and of being led to indulge in practices injurious to both body and mind.' As a result of all this, 'the young female returns from school to her home a hysterical, wayward, capricious girl; imbecile in mind, habits and pursuits; prone to hysteric paroxysms upon any unusual mental excitement'. Fashion played a role as well, with 'the waist (in despite of all warnings) … compressed into the most preposterously diminutive propor-tions', and the 'mammae are compressed (if beyond the fashionable magnitude) and irritated, and react upon the ovaria and uterus'.[45] The long-term consequence of such behaviour being 'moral insanity'. By this stage the:

> gentle, truthful, and self-denying woman has unaccountably become cunning, quar-relsome, selfish; piety has degenerated into hypocrisy, or even vice; and there is no regard for appearances, or for the feelings or interests of others, except in so far as they may minister to the vanity or selfishness of the patient.[46]

Laylock repeated himself in much the same terms the following year in the *Journal of Psychological Medicine & Mental Pathology*. Taking advantage of the conven-tional anonymity to review his own work in glowing terms, he added 'excessive devotion to needlework' to the list of dangerous and depraved habits of modern young womanhood.[47]

The cure to all of this was, of course, a good dose of galvanism. Laycock had no doubt that the 'remedy I have to recommend as prophylactic and curative in cases of this kind is the persevering and systematic application of electro-galvanism to the abdominal and pelvic regions'.[48] Galvanism acted as 'an aperient and emmenogogue' which had the effect of clearing the system, 'restoring the due action of the intestinal canal, and of the kidneys and ovaria' so that 'the morbid cerebral phenomena disappear, and the mind returns to its natural state'.[49] Golding Bird favoured electromagnetic treatment in such cases as well. It had the advantage, he thought, of curing the problem whether the patient was faking it or not. He recommended 'an uninterrupted current of an electro-magnetic machine' in such cases. 'If the patient simulates paralysis,' he explained, 'she can seldom resist the pain and surprise of the shock, and the previously rigid limb will instantly move.' If on the other hand, the hysteria was genuine, 'there are few curative remedies so important as the electro-magnetic current'.[50] He thought electricity useful in cases of hysterical retention of urine as well, having 'succeeded in curing this annoying symptom by passing a pretty strong current from the sacrum to the pubes'. Again, he suspected that 'the pain of the current and dread of its repetition have constituted the real elements of success in these cases'.[51]

In many ways for early Victorians, the most striking thing about electricity was its connection to the body. It was the stuff of life. Electricity offered the opportunity to heal the sick and revitalise the jaded. It also occupied rather dangerous cultural territory. The connotations with materialism and dangerous political radicalism had certainly not disappeared; neither had the overtones of charlatanry and shady showmanship. Electricity's proper place looked to be the marketplace – and none too respectable at that. But just as clearly, electricity seemed to offer a lifeline to ambitious doctors on the make. It provided a new tool with which they could try and forge a reputation for themselves. On the one hand, this meant that they needed to find ways of talking about and using electrical therapies that would distance them from their more unsavoury fellow galvanists. On the other, it meant they needed to reach out and persuade new audiences that they were people to trust with their bodies. Men like Thomas Laycock or Golding Bird were intent on turning electricity into an instrument of control. With it, they could offer themselves to society as guardians of the moral order. But for them, as for the likes of Halse or Leithead, electricity also seemed to offer an important tool for probing the relationship between human bodies and minds and the world around them. This was what Ada Lovelace had very much in mind as she set out for Broomfield and her encounter with Andrew Crosse. She was well aware, also, of the links that were being forged between electricity, hysteria and the essence of femininity, and what they might mean for her own condition and her experimental ambitions.

NINE

ADA'S LABORATORY

As far as Ada Lovelace was concerned, electricity was personal. By the time she visited Fyne Court in 1844 she was desperately ill. She complained earlier in the year that she had felt 'very queer of late' and suffered seizures in which she felt her head swelling, leaving behind 'most strange feelings in the brain & eyes'. Her doctor, Charles Locock, thought it was due to a sudden tension in her nervous system.[1] Locock fed her laudanum as a sedative. Ada's view was that 'Galvanization in *winter*, & sunbathing in *summer*' was the solution.[2] Electricity offered more than just a cure, though; it offered a way for her to understand her illness as well. Beyond that, she also clearly felt that electricity offered her a chance to turn her illness to an advantage. Her physical and mental condition meant that she could turn her body into a laboratory to investigate the relationship between mind and matter; between life and the universe. We can certainly assume that she was quite well informed about recent developments in the field. She did, after all, move in the right kinds of social circles. She frequented the Royal Institution and Charles Babbage's notorious soirées. She was acquainted with the men who mattered amongst the elite of English science. She would have had plenty of opportunities to become familiar with the fringes of electrical speculation too. Her correspondence certainly suggests a familiarity with even some of the most radical and ambitious galvanists, theorising about the nature of the relationship between electricity, mind, the body and the cosmos.

Ada took comfort in the fact that her body was:

> so *susceptible* that it is an *experimental laboratory* always about me, & inseparable from me. I walk about, not in a Snail-Shell, but in a *Molecular Laboratory*. This is a new view to take of one's physical frame; & amply compensates me for *all* the sufferings, had they been even greater.

She was 'simply the *instrument* for the divine purposes to act *on* & *thro*'; happening to be appropriate for that object'. She was going to be 'the Elijah of *Science*'.[3] She confided her plans to her mother and to Somerville's son Woronzow Greig, as well as to her husband. Before arriving at Fyne Court to embark on the next step of her grand plan of self-experimentation, Ada was anxious to assure Andrew Crosse that there was nothing impious about her project. 'Religion to me is science, and science is religion', she told him.[4] She also needed to warn Crosse about her illness. She told him that she was 'subject at times to dreadful physical sufferings. If such should come over me at Broomfield, I may keep to my room for a time. In that case all I require is to be *let alone.*'

It seems likely that some, at least, of Ada Lovelace's wilder speculations in the months leading up to her visit to Andrew Crosse were the result of the laudanum. It is clear from her correspondence that she was taking opium in significant quantities during this period and was aware that it was having an effect on her perceptions. She suggested that laudanum had a 'remarkable effect' on her eyes, 'seeming to *free* them, & to make them *open & cool*'. The drug made her 'so philosophical, & so takes off all *fretting* eagerness & anxieties. It appears to *harmonize* the whole constitution, to make each function act in a *just proportion*'.[5] This does not mean that her speculations should not be taken seriously, they were still efforts on her part to make sense of her predicament and turn it to good use. Ada was quite serious about the divine origins of her scientific ambitions; she wanted to use electricity as her road to salvation and in this respect treated her illness as a sign of divine grace. It was only because she was ill that she could consider turning herself into an electrical laboratory, and that made her illness and the use she hoped to make of it God's will. Indeed the language she used to describe her aspiration had as much in common with early Victorian evangelicalism as it did with radical materialism.[6]

Ada made no bones about the religious significance of her scientific endeavours when she initiated correspondence with Michael Faraday, begging for his assistance and putting herself forward as a potential assistant and collaborator. She had been '*vowed to the temple*', she told him – 'the Temple of *Truth, Nature, Science!*' She declared that she hoped 'to die the *High-Priestess* of God's works as manifested on this earth'.[7] Ada wanted to be admitted to 'intercourse & friendship' with Faraday. She wanted to be his disciple and was sure that 'any assistance & intercourse you may consent to honour me with, *would* pay full interest to *yourself* & for your *own* scientific & moral purposes in this world'. Ada suggested that Faraday might go through his *Experimental Researches* with her, '*showing* me the examples & experiments practically, as I study each paper'. She described her goal as 'the study of the *Nervous System, & its* relations with the more *occult* influences of nature'. For this she would need 'a masterly union of the highest abstract *analysis*, with most skilful & varied courses of *experimental & practical science*'. The result would be 'to bring the actions of the *nervous* & vital system within the domain of *mathematical* science, & possibly to discover some great vital law of *molecular action*, similar for the universe of *life*, to *gravitation* for the *sidereal* universe'. Ada thought it would be 'the work of a *life-time*'.[8]

Ada went on to quiz Faraday about his religious sentiments, declaring herself a Unitarian. It is extremely unlikely that Faraday would have replied to such a letter had it been written by anybody other than Lady Ada Lovelace, Byron's daughter. As it was, he replied a little over a week later. He informed her regretfully that his own illness and age (he was 53) made the sort of collaboration and discipleship that Ada wanted impossible. 'You have all the confidence of unbaulked health & youth both in body & mind; I am a labourer of many years' standing made daily to feel my wearing out,' he wrote.[9] He politely rejected her union of philosophy and religion. Faraday told Ada he was a member of the 'very small & despised sect of Christians known, if known at all, as *Sandemanians*'. Although of course he agreed that any study of the natural world could only glorify God, he 'did not think it at all necessary to tie the study of the natural sciences & religion together and in my intercourse with my fellow creatures that which is religious & that which is philosophical have ever been distinct things'.[10] This was a rebuff, and Ada surely recognised it as such, though their correspondence continued for several weeks. She tried to tempt Faraday with the suggestion that with his guidance she might 'make a *review & abstract*' of his work, 'for the Quarterly perhaps, or some such vehicle'.[11] Faraday was having none of it, although he remained polite and responded to her intimacies with rare intimacies of his own.

Despite the exchange of pleasantries and the playful tone of the letters, Ada clearly did not know Faraday very well. She would otherwise have been aware that Faraday had never adopted a disciple or a pupil in the way that Humphry Davy had adopted him – though he would come close with John Tyndall, his eventual successor at the Royal Institution. She might also have realised that Faraday would have very little sympathy for her grand vision of discovering an electrical theory of life – and even less for the way in which she proposed going about it – as was clear from his account of his experiments with the Adelaide Gallery's electric eels. Faraday certainly believed in self-help. He regarded himself, after all, as being a self-made man. This was what the mythology that already surrounded him claimed as well. The Duke of Somerset, as President of the Royal Institution, enthused to Charles Babbage that the 'story of Faraday is sure to make a great noise'. The astute duke was sharp enough to recognise a fairy tale in the making. There was 'something romantic and quite affecting in such a conjunction of Poverty and Passion for Science', he mused, 'and with this and his brilliant success he comes out as the Hero of Chemistry'.[12] It was a personally engineered and driven rise from rags to scientific riches, and the Royal Institution's president could see the reflecting glory on his august establishment, something with which Faraday was fully complicit. It suited him to trace his achievements to his own iron self-discipline and unique character, rather than to the years of tutelage with his master Humphry Davy.

Faraday had certainly set about the task of fashioning himself as natural philosopher for the gentry with zeal and determination. Even before he encountered Humphry Davy for the first time, Faraday was looking for a way into science. He became a member of the City Philosophical Society – a gathering of artisan scientific enthusiasts dedicated to self-help. At the Royal Institution he practised not just experimenting, but lecturing as well. He took elocution lessons and devoured Isaac Watts' *Improvement of the Mind*, ingesting Watts' strictures on learning by observation, reading, listening to lectures, conversation and meditation. Those who knew him noted Faraday's punctilious attention to detail and the care and discipline with which he prepared his experiments. His friend and future biographer John Tyndall knew just what made his predecessor at the Royal Institution tick. He was 'a man of excitable and fiery nature; but through high self-discipline he had converted the fire into a central glow and motive power of life, instead of permitting it to waste itself in useless passion'.[13] Discipline was at the heart of Faraday's success, and it was central to his view of natural philosophy and the art and limits of experiment as well. Wild leaps of intuition had no place in his epistemology. What mattered was discipline – and moreover the ability to discern the requisite qualities of discipline and restraint in others. Faraday would certainly not have seen much discipline and restraint in Ada and her ambitions.

In 1854 (two years after Ada's death), Faraday gave a lecture on mental education at the Royal Institution before Albert, the Prince Consort. This lecture was Faraday's manifesto of the knowable and unknowable. It started with some strictures on the limits of human knowledge that would have gone to the very heart of Ada's project if she had lived to hear them. Nothing he was about to say, Faraday insisted to his audience, had anything to do with human knowledge of the divine. 'Let no one suppose for a moment,' he warned, 'that the self-education I am about to commend in respect of the things of this life, extends to any considerations of the hope set before us, as if man by reasoning could find out God.'[14] Whilst there could be no incompatibility between God's truth and the truths arrived at by reason and experiment, any understanding of God's truth was derived from faith alone. There was an 'absolute distinction between religious and ordinary belief'. This was a strong statement to make in the face of prevailing orthodoxy. William Paley's natural theological argument held that the rational study of nature provides evidence for the existence of a Creator, and that the organisation of nature also held lessons for the rational mind about the attributes of God; and this remained the dominant view amongst the gentlemen of science.

The key to knowledge (about mundane matters at any rate) was judgement. Knowledge was a matter of properly judging how to interpret the world and – crucially – judging whose claims could and could not be trusted. When we were fooled by our senses, it was not because our senses were faulty, but because we failed to properly judge what we had seen. Similarly, we were led astray by failing to judge the difference between authorities (those who shared Faraday's self-discipline) and charlatans. Charlatans were themselves more often guilty of self-deception – a failure to properly exercise their judgement – than intentional fraud. There were 'multitudes who think themselves competent to decide, after the most cursory observation, upon the cause of this or that event'; they were almost invariably wrong. The answer was self-education and the key to self-education was self-disci-pline. Self-education depended on humility and a willingness to accept correction. It also depended on having the strength of character to resist temptation. In fact, this was the key as far as Faraday was concerned: 'that point of self-education which consists in teaching the mid to resist its desires and inclinations, until they are proved to be right, is the most important of all, not only in things of natural philosophy, but in every department of daily life.' Knowledge was a matter of proportionate judgement – of being aware of the strength of arguments and evidence for one proposition or another. It was also a matter of judging what kinds of investigations were worth following and which were not.

Faraday had some very specific targets in mind in his lecture on mental edu-cation. His main target was the new-fangled craze for table-turning that swept the country during the 1850s. The phenomenon itself was straightforward. When a group of participants gathered together around a table, all placing their hands on the table's surface and pressing down, the table moved to right or left. Some commentators suggested that this demonstrated human electricity in action. Table-turning séances were soon all the rage in fashionable circles. As far as Faraday was concerned, this was an example of the perils of undisciplined minds. People fell for table-turning both because they failed to properly interpret the evidence of their senses and because they allowed themselves to be seduced by a desire for the spectacular and sensational. Their wish to see the order of nature suspended warped their judgement. Typically, he designed an experiment to demonstrate his point. His apparatus showed that the participants' hands moved before the table did – so the hands moved the table rather than the table moving the hands. It was an instance of what Faraday politely termed a 'quasi involuntary muscular action'.[15] The par-ticipants literally did not know what their hands were doing and needed Faraday's apparatus to tell them what was actually taking place. They lacked the appropri-ate mental discipline and therefore experienced what they wanted to experience, rather than what was really going on.

But table-turning was just one example of a more common philosophical malaise that derived from a failure to understand how knowledge was to be produced and evaluated. Faraday had mesmerism, phrenology and what he regarded as the wilder electrical speculations in his sights as well. Mesmerism and phrenology both had their roots in late Enlightenment Vienna. Franz Anton Mesmer had developed the theory and practice of animal magnetism, as he called it, as a way of manipulating the human body's magnetic fluids for therapeutic purposes. According to Mesmer, illnesses resulted from imbalances in the body's supply of magnetic fluid. Using a variety of techniques he claimed to be able to redirect the flow of magnetism through the body, thus restoring the natural balance essential for physical and mental health. The science of phrenology had its origins in the research of the Viennese physician Franz Josef Gall, which looked into the relationship between the size and shape of the skull and a particular individual's mental character. Gall argued that the brain was divided into a number of different organs and that each of these organs represented a specific mental faculty. The size of each of these organs offered a measure of the strength of that faculty in the individual and could be measured by examining the shape of the skull. Both mesmerism and phrenology enjoyed widespread popularity across Europe at the end of the eighteenth and beginning of the nineteenth centuries. Both sciences were regarded by opponents and proponents alike as offering materialist theories of the operation of the mind.[16]

Following a decline in its popularity during the war with France and its aftermath, there was a resurgence of interest in mesmerism in Britain from the 1830s and throughout the early Victorian period.[17] Public mesmeric performances and lectures flourished in London and across the country. Mesmeric performers advertised themselves as therapists and offered public demonstrations of mesmeric phenomena. The radical bastion of London University became a centre for mesmeric research too. Dionysius Lardner, who had briefly been Professor of Natural Philosophy at London University before resigning in 1831, was an enthusiast for mesmerism and an ardent promoter of its materialist implications. John Elliotson, Professor of the Principles and Practice of Medicine at London University, also shared Lardner's materialist leanings. In 1837 he conducted a series of public mesmeric experiments on the wards at University College Hospital using two servant girls – the O'Key sisters – as his main subjects. The good and the great amongst London's medical and scientific practitioners – including Faraday himself – were invited along to see the O'Keys' performances until their notoriety forced Elliotson's resignation from both university and hospital a year later. Undaunted, and financed by the profits from his lucrative medical practice, Elliotson continued to proselytise for mesmerism throughout the 1840s, establishing the *Zoist*, a magazine of mesmerism and phrenology in 1843, and founding the London Mesmeric Infirmary in 1849.

Ada Lovelace was certainly interested in mesmerism. In fact, in the months running up to her visit to Andrew Crosse at Fyne Court, her correspondence demonstrates a particular fascination with the latest mesmeric scandals titillating fashionable and intellectual circles: the case of Harriet Martineau. In 1844 Martineau was probably one of England's most prolific writers. She was the author of tracts on political economy, a number of novels and numerous contributions to the *Westminster Review* and other journals. Her *Life in the Sickroom*, published in 1844, offered a detailed account of her recent illness and of her cure by mesmerism. Ada avidly discussed Martineau's book and the subsequent letters in the *Athenaeum* in which Martineau described her mesmeric experiences with family and acquaintances. Ada recorded that her physician, Charles Locock, thought there was 'nothing … *at all absurd*' about Martineau's account of her experiences with mesmerism, and that '*states of health* both *had* been, & *could* be benefited by its measures', though he doubted that 'any *organic* disease' could be cured. Ada told her mother that she had been 'thinking much about the Mesmerism in Miss Martineau's case' in relation to her own experiences and illness.[18] Many gentlemen of science, on the other hand

– including Harriet Martineau's own brother-in-law, Thomas M. Greenhow – vili-
fied Martineau for presuming to affirm the value of her own personal experience
and testimony against the weight of authoritative opinion.[19]

Supporters of phrenology also found themselves at odds with the claims and
values of the gentlemen of science. Phrenology's promoters emphasised its acces-
sibility as one of its most important features. Phrenology did not require years of
training and the development of difficult skills. All a potential practitioner needed
was a phrenological chart identifying the different organs of the brain and their
function. A little practice at feeling the bumps on the head that showed the location of the various faculties and their relative sizes was then all that was needed to
become a phrenologist. In the hands of advocates such as George Combe, phrenology also became associated with a culture of self-help and self-improvement.
Combe's bestselling *Constitution of Man* made phrenology into a practical moral
philosophy. It offered a road not only to individual improvement, but to the moral
improvement of society as a whole. It offered its adherents a set of tools that
would allow them to understand themselves in relation to the world around them.
Through phrenology they could gain an understanding of their moral strengths
and weakness and act accordingly. Phrenology also seemed to offer a powerfully
egalitarian message. If character and ability were functions of the physical structure
of the brain, then there was no real barrier to social aspiration. A farm labourer or
factory worker's sons, if their brains were appropriately organised, were as capable
of leading the country as the sons of the gentry.

Where Faraday regarded these kinds of popular practices with a jaundiced eye, it
is clear that Ada found them fascinating. Both phrenology and mesmerism seemed
to offer potential models for her own researches. They straddled the boundary
between the inner and outer worlds in the same sort of way. This kind of boundary breaking did not fit well with Faraday's ideas about knowledge. Neither did
the egalitarian ethos that underlay such views. In fact, the mesmerists' and phre-
nologists' claim that their practices were open to anybody was exactly the kind
of thing that made Faraday suspicious. Faraday made it quite clear in his lecture
on mental education that the 'multitudes' unprepared to undergo the rigours of
self-disciplined self-education had little useful to say about matters of fact. In this,
at least, Faraday's opinion was shared by most other gentlemen of science. Natural
philosophy was the preserve of the few rather than the many. For someone like
Ada's mentor Charles Babbage, science was to be regarded as a vocation that was
accessible to only a privileged few. There was a strict division to be drawn between
those who had the mental capacity to practise natural philosophy and those who
formed the audience for their work. It was a division that could be seen in action at
any gathering of the British Association for the Advancement of Science, or at one
of the Royal Institution's Friday evening discourses. And there can have been little
question in the end on which side of the boundary Ada stood as far as Babbage and
Faraday were concerned.

There was a great deal at stake in these epistemological niceties. These kinds of views about who should and should not be regarded as trustworthy interpreters of the natural order lay behind the campaign that Babbage and his allies had mounted throughout the 1820s for the reform of the Royal Society. This campaign had kicked off with the death of Sir Joseph Banks, the old president who had ruled the Society with a fist of iron for more than forty years. Babbage and his cronies regarded the Royal Society's old guard as amateurs and dilettantes. They had to be swept away to make room for a new generation of vocationally minded gentlemen like himself, who possessed the right kind of mental discipline and managerial skills for ruling the world of science. [20] Following the failure of his campaign to get his friend John Herschel elected as President of the Royal Society rather than the Duke of Sussex in 1830, Babbage retaliated by publishing *Reflexions on the Decline of Science in England*. He lambasted the scientific establishment for its corruption and nepotism, casting an envious eye over the Channel, at France's system of state sponsored savants. Knowledge was best produced in the same way that factories produced commodities, according to Babbage, with a strict division of mental labour. Gentlemen natural philosophers like him, trained and self-disciplined, were the scientific world's equivalents of the factory manager. [21]

Babbage's Calculating Engine embodied his politics of knowledge. His views about the division of mental labour and the hierarchy of knowledge production were built into the machine's gears and cogwheels. The Calculating Engine was designed originally to take the drudgery out of calculation. Babbage recalled sitting half-asleep over a table of logarithms in his rooms in Cambridge when he was interrupted by a fellow student. He had been day dreaming about finding some way in which those dreary logarithmic tables 'might be calculated by machinery'. [22] The Calculating Engine was his answer. It was designed to replace the labour of arithmetical calculation in exactly the same way that a mechanical loom was designed to replace the labour of hand-weaving. One of the things that the engine did was make the processes of calculation open and transparent to inspection. For Babbage and his intellectual ally John Herschel, this openness was an important requirement for the production of scientific knowledge. As Herschel put it, knowledge needed:

> its elements made accessible to all, were it only that they may be the more thoroughly examined into … it should be divested, as far as possible, of artificial difficulties, and stripped of all such technicalities as tend to place it in the light of a craft and a mystery, inaccessible without a kind of apprenticeship. [23]

It needed to be like this so that it could be managed.

Babbage argued that his engine was not just a good model for human knowledge; it was a model of the Divine mind as well. In his riposte to the Royal Society sponsored Bridgewater Treatises extolling natural theology, Babbage in his (unsanctioned) *Ninth Bridgewater Treatise* used the Calculating Engine as his exemplar. God had designed the universe so that it embodied exactly the kinds of principles that were embodied in the Analytical Engine – the latest version of Babbage's invention. The Analytical Engine exhibited the basic elements of intelligence: memory and foresight. It remembered what it had done in the past and could predict the form of future calculations. God had embedded the same kinds of characteristics in the operations of the universe as well. The Analytical Engine operated by means of punched cards that could be fed into the machine, conveying instructions about what calculations to make. The cards might, for example, tell the machine to calculate according to a particular formula a set number of times, and then to switch to a different calculation. According to Babbage, this was also how God had built natural law into the very fabric of the universe.[24] The Analytical Engine embodied Babbage's view of the production of knowledge – just as the universe embodied God's law – as an ordered, hierarchical activity that needed to be reduced to mechanical simplicity so that it could easily overseen and managed by disciplined gentlemen such as himself.

Babbage would have had as little sympathy as Faraday, therefore, for Ada's project to turn her own body into an electrical laboratory. Her scheme fitted neither of their views about how knowledge should be produced and scrutinised, or what kind of person the producer of knowledge should be. Just as Harriet Martineau argued that her personal experience of illness and the value of mesmerism gave her a privileged insight, so Ada Lovelace felt that her sick body gave her a particular and personal insight into the relationship between electricity, her body, her mind and the universe. Whilst this kind of claim fitted poorly with the view of natural philosophy and its practitioners advocated by the likes of Babbage or Faraday, it had resonances with the epistemologies put forwards by other electricians, such as some of the members of the London Electrical Society, or promoters of mesmerism or phrenology. Ada's experiment with herself is revealing because she sat on the fissures that fractured early Victorian scientific culture. One might even say that only those who had no other choice – women and the disenfranchised who were excluded from the inner world of Victorian natural philosophy – ended up turning their own bodies into laboratories. Her epistemological project and the personal politics of knowledge that combined with it went to the heart of who could and could not be considered as legitimate producers of knowledge. By using her body and her illness in this way she was subverting the usual early Victorian view of the relationship between body and mind in making knowledge. She was making a virtue of her bodily dependence.

Ada tried to take full advantage of her visit to Fyne Court to consolidate her knowledge and her practical experimental skills. It is clear, nonetheless, that she had mixed feelings about its success. The world of experiment as she encountered it in Andrew Crosse's laboratory was clearly not as she expected it. She complained that there 'is in Crosse the most *utter* lack of *system* even in his Science'. There is little in her correspondence about any experimental work she might have done there.

In fact, she worried that she had 'quite a difficulty to get him to show me what I want. *Nothing* is ever *ready* All chaos & chance.'[25] She had clearly found plenty to talk about, on the other hand, and Crosse's son turned out to be full of information about the latest developments in German science. Ada left Fyne Court feeling that her 'visit to Broomfield seems to have been very opportune & to have laid the foundations for much that is wanted for me at this epoch of my progress'.[26] Despite this report, there is very little evidence to suggest that she continued with her electrical enthusiasm following the visit. Ada's project to turn her own body into a laboratory seems to have ended up going nowhere.

Ada Lovelace's visit to Andrew Crosse and Fyne Court provides an insight into both the broad culture of electricity at the beginning of Victoria's reign and the intimate relationship that many perceived between electricity and the body. Lovelace and Crosse inhabited an experimental culture in which it made perfect sense to carry out electrical experiments in the hope of reaching a new understanding of the mysteries of life and consciousness. In this world, Ada's ambition to turn her body into a laboratory was both sensible and achievable. Ada Lovelace's experiences also show, however, how, by the early years of Victoria's reign, the experimental electrical world was well on the way to becoming marginalised within the dominant metropolitan culture of the gentlemen of science. These gentlemen of science were desperate to cut the cord tying experiment to political radicalism. They wanted their science to be divorced from any such unsavoury connections that had tainted it in the past. There was no room, therefore, in this newly disciplined culture of Victorian experiment for speculation about electricity and life. To find out about it and experience it, even somebody as intellectually curious, well-connected and well-informed as Byron's daughter had to travel to an obscure Somerset estate and Andrew Crosse's laboratory.

PART III

CONSTANCE PHIPPS

TEN

LADY CONSTANCE'S PAIN

Constance Phipps was ill and confined to her bed. She was suffering from head-aches and probably complained of sensations of numbness in her arms and legs. The correspondence between her father and one of her medical practitioners, discussing the details of her treatment, does not specify exactly what was wrong with her, but from it we can learn several other things. The letters are undated but other evidence indicates that they were almost certainly written sometime between 1863 and 1871. Constance at this time would have been a young woman, somewhere between the ages of 11 and 19 – what we would call a teenager, though the Victorians had no such term in their vocabulary. It is further clear from the correspondence that a portion of Constance's treatment was being administered by her father, overseen by at least two other medical practitioners. One of these practitioners was Harry Lobb, who by the beginning of the 1860s was starting to forge a formidable reputa-tion as an advocate of medical electricity. He specialised in the electrical treatment of neuralgia and nervous complaints more generally. Constance Phipps was at just the right age and was a member of the appropriate social class and gender to be regarded as particularly at risk of precisely these kinds of ailments.

    The position in which Constance found herself was therefore not particularly unusual. By the middle of the nineteenth century, electricity was broadly accepted as an effective form of medical treatment for a wide range of nervous disorders by many medical practitioners, although its use as a therapy was also considered to be tantamount to outright quackery by many others. Self-proclaimed specialists in nervous diseases, such as Harry Lobb or Thomas Laycock, argued that young ado-lescent women like Constance were particularly prone to these kinds of conditions as a result of their sex and upbringing; they were at the mercy of their reproductive organs. Young women of the middle and upper classes, cosseted and spoilt by over-indulgent parents and subjected to the whims of fashionable society, were almost inevitably given to hysteria – or so argued those men whose interests lay in promising ways of restoring equilibrium. Electrical treatments were offered to anxious paterfamilias as a way of bringing wayward family members back under paternal control.[1] It is not clear how common it was for the anxious fathers to carry out the electrical treatment with his own hands, as seems to have happened in Constance's case, but advocates of electrotherapy quite often emphasised that this was a regimen that could be administered safely by a family member or servant under the appropriate medical direction. In many respects then, it seems reasonable to say that the cast of characters in this particular little drama were, if not typical, then at least not greatly out of the ordinary.

Paradoxically, the member of this cast about whom least is known is the leading lady. Very little can be gleaned about the short life of Lady Constance Mary Phipps. Her voice is heard only briefly, faintly – and offstage – through her father's correspondence with Harry Lobb. We know she was born in December 1852 in the parish of St George's Hanover Square in London. Her father, George Augustus Constantine Phipps, was the only son and heir of the first Marquess of Normanby. Her mother, Laura Russell, was the daughter of Robert Russell, a naval captain and himself a scion of aristocracy. She was one of seven children and the youngest of three daughters. Constance herself appears just three times in the Victorian state's meticulous official records. Her birth and death – just thirty-one years apart – are recorded. There is only one other glimpse of her, in the census record for 1881, two years before her death. She was staying with her sister Katherine and her husband Francis Egerton, the third Earl of Ellesmere. She was the only adult family member recorded in the house on the day of the census (though there were plenty of servants, of course). As the resident maiden aunt, she had presumably been left in charge of the Egertons' large brood of children. She was therefore almost certainly unmarried. Whilst this was not uncommon amongst the mid-Victorian upper classes, it is unusual enough to merit comment. Her single state, along with her adolescent invalidism and early death, all hint at a history of illness.

Constance came from distinguished stock. The Phipps were descended from Sir Constantine Phipps, Lord Chancellor of Ireland at the beginning of the eighteenth century. Through his son William's marriage to Catherine Annesley, daughter of the Earl of Anglesey, a touch of royal (if illegitimate) blood was added to the dynasty as well. By the middle of the eighteenth century, the family had acquired an Irish peerage as earls of Mulgrave and later acquired an English peerage of the same title as well. Constance's great-grandfather was elevated as Viscount Normanby in 1812 and her grandfather became the first Marquess of Normanby in 1838. The Phipps had gained their position as Tory placemen. Henry (the great-grandfather) was one of Pitt the Younger's loyal henchmen and an inveterate enemy of the 'abominable doctrines of equality'. Her grandfather, however, turned coat to join the Whigs, became Home Secretary under Lord Melbourne, and was British ambassador in Paris in 1848 during the revolution that brought Louis Napoleon to power, before falling out with Palmerston. He was on the liberal wing of his party – a campaigner for Catholic emancipation (the issue on which he fell out with the family's traditional politics), an opponent of slavery and a staunch advocate of political reform. He penned romantic novels and political sketches. Constance was born, therefore, into a dynasty of politicians, diplomats, courtiers and intellectuals.

Her father George Phipps, who inherited the title in 1863, was a career courtier and diplomat in the family tradition. In 1847 he commenced his political career by becoming the Whig Member of Parliament for Scarborough – the local constituency for the family seat near Whitby, and presumably in the family's pocket. Four years later, after being appointed Comptroller of the Royal Household, he lost the seat but regained it the following year. For much of the 1850s Phipps was a parliamentary whip for the Aberdeen and Palmerston governments, and in 1858 was appointed lieutenant-governor of Nova Scotia. He remained in Canada until 1863 when his father's death brought him back to England as the second Marquess of Normanby. Here he stayed until 1871, supporting the Liberal party in the House of Lords under Lord John Russell and Gladstone – during which time the letters to Harry Lobb must have been written. In April 1871 he was appointed governor of Queensland in Australia, after telling Gladstone that he wanted to spend the rest of his career in the colonial service. It was a lucrative business. The Normanby estates were comparatively small and as a result the Phippses lacked an income to match their social and political ambitions. A colonial governor, on the other hand, could expect a salary of £10,000 a year. The governorship of New Zealand followed in 1874 and in 1879 he became governor of the state of Victoria in Australia. He stayed there until his wife's death in 1884, after which he retired from the colonial service. Constance's parents were therefore probably both in Australia when she died in October 1883.

We may speculate that it was through his connections at court that Lord Normanby came into contact with the medical electrician Harry Lobb. Whilst still the Earl of Mulgrove he had been appointed Comptroller of the Household to Victoria's court. In this capacity, the earl was responsible for overseeing a variety of the court's ceremonial proceedings. A few years later, in 1853, Mulgrave was appointed Treasurer of the Household, with responsibility for aspects of the royal finances – those that came from the public purse, at least. Both of these appointments whilst in principle the monarch's gift, were in practice political – they were made by the prime minister. Nevertheless, a glance through the Court Circular pages of The Times suggests that Constance's father was a frequent attendant at court throughout his period in office. There he might well have encountered the queen's physician and accoucheur, Charles Locock (whom we have already encountered as Ada Lovelace's doctor). Locock was responsible for overseeing the queen's pregnancies. He was an enthusiast for new medical technologies, and was one of the first doctors to use chloroform as an aid to delivery, using Victoria as a guinea pig to demonstrate its efficiency. Locock was therefore a keen advocate of medical electricity (though he does not seem to have published anything on the topic himself) – and used his influential position at Victoria's court to promote the therapy. One of his protégés was the rising star of electrotherapeutics during the late 1850s: Harry Lobb. It seems likely that it was Locock who brought Lobb to the Earl of Mulgrave's attention.

Lobb was himself the latest member of a prosperous medical dynasty. His father, William Lobb, was a well-known and successful London doctor who had carved out a large and lucrative metropolitan practice for himself by the 1840s. Harry studied medicine at St Bartholomew's Hospital and passed his examinations to become a member of the Royal College of Surgeons and a licentiate of the Society of Apothecaries in 1850. According to his own recollections, Harry attended lectures at Guy's Hospital as well.[2] If he attended lectures at Guy's during the second half of the 1840s he may well have come across Golding Bird's introductory lectures on natural philosophy. Here, then, he might have encountered the possibilities of medical electricity for the first time. Lobb senior, however, was clearly something of a curmudgeon and less than enthusiastic about supporting his son's budding medical career. Even before Harry completed his studies his father was indulging in *Schadenfreude*. 'I do not think you would ever work and do as I have done, and without such devotion there is no chance of success,' he told his son.[3] In the end, Lobb signed up as ship's surgeon with the East India Company and spent two years in the east aboard the *Herefordshire* before returning to London.

Back home after his stint at sea, Lobb set about establishing a practice of his own in London, working from his home at 63 Gloucester Terrace, a recently completed street of handsome terraced houses in newly fashionable Bayswater.[4] At some stage during the latter half of the 1850s, he somehow succeeded in bringing himself to the attention of Charles Locock. Whilst Lobb was still based at Gloucester Terrace, Locock was recommending his services to potential customers, suggesting to one sufferer that the 'effect of galvanism is often very remarkable in such cases of headache and it would be worth while to send to Mr. Lobb – 63 – Gloucester Terrace, Hyde Park, saying in your note what the case is'. He told another that 'I wish Mrs. Brookfield would see Mr. Lobb, of 63 Gloucester Terrace Hyde Park, who has had great success in treating causes of partial paralysis with *proper* Galvanism – much harm being done often by using it wrongly or in improper degrees'.[5] In 1858 Lobb published his first book, *On Some of the More Obscure Forms of Nervous Affections*. His next book, *On the Curative Treatment of Paralysis and Neuralgia, and Other Affections of the Nervous System with the Aid of Galvanism*, published the following year was dedicated to Locock as 'a testimony of the valuable assistance afforded by him in the establishment of galvanism as a *recognised* agent in the treatment of disease'.[6] By now he was established at the even more prestigious address of 70 Brook Street, off Hanover Square.

Locock's patronage was clearly paying dividends. Lobb was attracting wealthy and influential patients as well as making a public reputation for himself with his books and other publications. In 1861 he opened his own hospital, the London Galvanic Hospital, with a high-rent location on Mortimer Street, off Cavendish Square, and boasting a long list of aristocratic patrons.[7] The hospital was established 'for the treatment, with the aid of Electricity, of all forms of Nervous and Muscular Diseases, for which this force is particularly adapted'. Its promoters hoped to offer cures for those 'helpless, abandoned, and so-called incurable cases' that other hospitals could not help, as well as 'forming the nucleus for a school of electro-physiology and therapeutics'.[8] An editorial comment in the *Electrician*, penned presumably by the journal's indefatigable editor, the telegraph engineer and spiritualist sympathiser Desmond FitzGerald, came out in strong support of Lobb, the hospital and the practice of medical electricity. The new establishment would be 'the means of affording an asylum for this science, a home from whence all our English discoveries will emanate', FitzGerald wrote. Scholars and sufferers alike would rally to the cause 'when they see a force, resembling that force upon which life depends, enabling them to cure all manner of diseases, which hitherto had baffled all remedies, of whatever nature they might be'.[9]

Neither Lobb's nor FitzGerald's optimistic prophecies were fulfilled, however. The London Galvanic Hospital went to the wall only a few years after its foundation. By the 1860s Charles Locock had retired from active medical practice and his support was clearly proving progressively less useful. Without this active patronage, Lobb's network of aristocratic patients and patrons seems to have gradually bled away. Lobb worked hard to maintain his standing, nevertheless; he published prolifically and delivered public lectures on his work. The supportive Desmond FitzGerald at the *Electrician* provided one potential lifeline by printing a series of his lectures in the journal, with enthusiastic encomiums. And Lobb read papers at the prestigious Harveian Society of London several times. In December 1859, for example, he discussed 'the pathology and treatment of idiopathic peripheral neuralgia'. In 1863 he spoke on 'the uses and value of galvanism and electricity in general practice' to his fellow practitioners.[10] It was, nonetheless, Lobb's publishing activities and his diligent attempts at self-promotion that eventually brought about his downfall. In 1874 Lobb took out a series of one-column advertisements in *The Times*, puffing his books and his electrical practice. This was not the kind of activity that went down well with fellow medical gentlemen, and Lobb was hauled over the coals and threatened with the loss of his membership by the Royal College of Surgeons, despite his avowal that 'the sense of the profession, that is to say those who have written books', was on his side.[11] The Harveian Society tried to expel him but failed to get the required two-thirds majority. When the Society's president and several others resigned as a result, Lobb had little choice but to submit his own resignation.[12] From that point on, his reputation in tatters, Lobb largely disappears from the historians' view.

This series of mini-biographies is interesting because it suggests something important about how things really worked in the underground world of mid-Victorian electricity. Who one knew mattered at least as much as what one knew. Lady Constance and Harry Lobb were linked together by networks of patronage and mutual interest that crossed boundaries of family, profession and science. Lobb was the doctor who oversaw her treatment because his patron, Charles Locock, had a position at Victoria's court that brought him into contact with Constance's father. It would have made perfect sense for Lord Normanby to turn to someone like Locock (who was, after all, well known as a specialist in female complaints) when looking for a trustworthy man to treat his ailing daughter. Trust was central to the relationship between men of science and their audiences, and to the relationship between doctors and their patients (or at least the people who paid them for their services). The networks through which these connections of mutual trust travelled were informal and vey often largely invisible. They depended on who knew whom, on social status, on convincing public performances and on proper and appropriate behaviour. Lobb's career eventually came unstuck over exactly that last issue. Issuing advertisements was the sign of a tradesman, not a gentleman. And upper-class clients and audiences expected their doctors and natural philosophers to at least present the appearance of gentlemen.

Given this background, how can we make sense of what took place in Lady Constance's bedroom? From the brief correspondence between Normanby and Lobb it seems that her father was treating Constance with what he described as a 'battery'. Indeed in two of the letters he complains that the 'battery' has broken. In one, he tells Lobb that the 'Battery is worn out & is of no further use'. In another, Normanby reports: 'I am sorry to say that the tiresome Battery has again worn out & no current can be obtained so that I must ask you to send me down another the chain has given way just the same as the last, several of the links having gone.' He discusses Constance's condition, telling Lobb that 'she has made a wonderful improvement since you saw her. She will be in London in about three weeks but if you wish her to go on with the Galvanism in the meantime I shall be much obliged to you if you will send me down a new battery.' It is clear that on at least one occasion Lobb had travelled to the Normanby family seat at Mulgrave Castle near Whitby. Normanby was specifically using the battery to pass currents of electricity through the affected parts of Constance's body – her head and her hand are mentioned in particular.[13]

The battery Lord Normanby was using was a Pulvermacher chain. These chains (or belts), invented by J.L. Pulvermacher, had become popular as medical electrical devices during the 1840s – and were to remain popular for much of the century.[14] They were simple to use and versatile, and their users could adapt them for a variety of ailments and injuries. A single element in a Pulvermacher chain in its simplest form, consisted of a coil of:

> thin zinc wire ... wound round a small wooden rod; one end of the wire entering the wood, the other terminating in a loop; at the opposite end to the zinc loop there

is a loop of thinner copper wire, which winds round the rod in the interspaces of
the zinc wire and pierces the wood at the opposite end.

This made a 'perfect miniature galvanic battery with a positive and a negative
pole'.[15] Any number of these elements could then be strung together to form a
chain whose electrical power varied according to the number of elements used. It
was activated simply by being soaked in a solution of vinegar and water, and was a
relatively foolproof device. Unlike many of the galvanic nostrums marketed from
the 1840s onwards, it also produced a fairly reliable flow of electricity. In principle,
rather little skill was needed to use a Pulvermacher chain, though Lord Normanby's
constant complaints about broken batteries suggest that some modicum of skill was
indeed essential.

The basic apparatus could be adapted in a whole range of ways in order to supply
a dose of electricity to the appropriate part of the body. If the chain was to be used
to pass a current of electricity through the head, for example:

> the white buckle A of the chain is placed on the forehead just below the hair, where
> it is attached by a ribbon surrounding the head and fixing itself to the two hooks of
> that buckle. The chain passes over the centre of the cranium, and ends at the nape,
> where the yellow buckle B, which forms its termination, is attached by a ribbon
> wound round the neck.[16]

Similarly, a current might be passed along the patient's arm by attaching one end
of the chain to a ribbon tied around the wrist, and the other to a ribbon tied
around the shoulder. Pulvermacher's pamphlets emphasised how important it was
that the 'force of the current must be adapted to the age, temperament, and nervous
excitement of the individual'.[17] As well as draping the chain over various parts of
the body, its users could simply hold both ends in order to pass a current through
them or sit with their feet in a basin of water holding one end of the chain whilst
the other end dangled into the water. They could even take an electrified bath with
one end of the chain attached to the metal bathtub whilst they held on to the other
end. Pulvermacher offered additional items, such as the 'Voltameter, or apparatus
for decomposing water' that could be used to measure the amount of electricity
passing through the body.[18] Lobb had certainly given Normanby one of these as he
mentions using it in one of the letters.

Lobb was a great advocate of Pulvermacher's chains as a convenient and useful
way of applying electricity to the body. It is therfore hardly surprising that he recommended (and presumably supplied) them to Normanby to use on Constance.
His early books contain detailed descriptions of Pulvermacher's apparatus – including some that he had had specially constructed. Case studies described treatments
of neuralgia using 'a 40–link chain from the nape of the neck, with the aid of different sized and shaped conductors, to the painful spots' or 'a 60–link Pulvermacher

battery, from the nape of the neck to the painful part'.[19] He also published articles in the *Lancet* on his successes using Pulvermacher's chain.[20] From the case studies and Normanby's own passing references to what he was doing, it is possible to reconstruct a general sense of the treatment being offered to Constance. In all probability the chain was indeed being used to pass a current of electricity through her head and to deliver a current along one or both of her arms. Lobb's descriptions of the kind of treatments he typically prescribed to his patients indicate that there was nothing unusual about Constance's treatment. It was routine; the only thing that makes it stand out at all is that the correspondence between Lobb and Constance's father provides a brief and tantalising glimpse of the realities of the sick room that usually lie hidden behind the doctor's published case notes. There is a hint of Constance as a human being: we know that 'she worries a great deal' about her treatment and condition, for example.

If the treatment that Lobb offered Constance represented nothing out of the ordinary by his own criteria, there is nothing to suggest that he was doing anything particularly unusual by the standards of his fellow electrotherapists either. By the 1860s electricity had achieved a degree of acceptance as a form of medical treatment. Although it was still a marginal practice, electricity had begun to lose some of the old radical and materialist connotations that had made it less than respectable a decade or two earlier. Practitioners who aimed for professional respectability continued to complain about the difficulty of being taken seriously by their fellow doctors, but it was the unacceptable whiff of the marketplace, rather than of the barricades, that electricity carried now. There is certainly every indication that someone like Lobb could maintain a large and prosperous practice as a specialist in electrical treatments. All the indications are that medical electricity was holding its own and that it had an expanding group of enthusiastic and dedicated advocates. Medical electricity also clearly benefited from the trend towards hospital specialisation that developed during the second half of the nineteenth century. Ambitious doctors unhappy with the restrictions of the large metropolitan general hospitals (or unable to find a place in them) branched out and established their own medical institutions devoted to the treatment of particular diseases, particular parts of the body or particular and idiosyncratic groups of patients like women, children, or the working classes.

Julius Althaus and Herbert Tibbits represent good examples of men who made electricity their medical specialism in this way. By the 1870s both held positions at, respectively, the Hospital for Diseases of the Nervous System in Marylebone and the West-end Hospital for Diseases of the Nervous System, Paralysis and Epilepsy on Welbeck Street. Tibbits had also been the medical officer in charge of electrical treatment at another of the new, specialist establishments, the Hospital for Sick Children on Great Ormond Street. Again, it is not surprising that medical electricity should find a therapeutic niche in these particular specialist hospitals. Practitioners who wanted to make electricity respectable usually did insist (unlike the quacks who maintained it was a panacea that would cure all ills) that its use was limited to specific cases and kinds of nervous disorders. Tibbits described his own institution as 'the pioneer and outpost, so to say, in this metropolis of the scientific and methodical application of electricity to the alleviation and removal of disease'.[21] Althaus and Tibbits were amongst the most prolific publishers of medical electrical texts throughout the 1860s and 1870s. What Althaus would have made of Lobb's delegation of Constance's treatment to her father is unclear. His *Treatise on Medical Electricity* (which went through four editions between 1859 and the end of the century) insisted that electricity 'was not one of those remedies which, if they do no good, can do no harm; but on the contrary, it may, in the hands of an inexperienced operator, do a great deal of mischief'.[22] More pragmatically, Tibbits acknowledged that in practice 'instructing *some one attendant of the patient* how to carry out the treatment, making her do this a few times in our presence, *and looking sharply after her afterwards*' was the best that could be managed.[23] Change the sex, and maybe that was what Lobb was trying to do with the Marquess of Normanby.

Not everyone welcomed the new specialist hospitals with open arms. Lobb's establishment of the London Galvanic Hospital was greeted with a torrent of vituperation from the editor of the *Lancet*. If the *Lancet* had started out as a brave mouthpiece of medical radicalism in the 1820s, it had become distinctly crusty by the 1860s. What possible use to anyone was a hospital devoted to treatment by electricity? One might as well establish 'a Quinine Hospital, an Hospital for Treatment by Cod-liver Oil, by the Hypophosphites, or by the Excrement of Boa-Constrictors', the editorial spluttered. Lobb was betraying his fellow practitioners and his professional training by indulging in such grandstanding; 'If men possessing the elements of respectability contained in a College diploma lend themselves to the foundation of such institutions, how can we wonder that quacks build up side by side their anatomical museums and their syphilitic cabinets?'[24] Anatomical museums and syphilitic cabinets were regarded as little more than thinly veiled pornography in respectable circles; this was getting pretty close to calling Lobb a pimp. Lobb could only respond by reiterating his qualifications, 'five years of practical study', waving his list of prestigious patrons and swearing that he would shut the galvanic hospital's doors in a second if one of the established hospitals would offer him a post and the resources to practise medical electricity.[25]

It is important to understand just what the *Lancet* thought was objectionable here. The problem, as far as they were concerned, was not electricity, but the hospital. There was nothing wrong with medical electricity in the right circumstances and in the right hands, far from it in their view. Medical electricity was already well established in its right and appropriate place within a respectable hospital practice, they averred. It was only making it into a specialism that shaded into dangerous and irresponsible quackery. This raises the question of just where the right and appropriate place for electricity – and bodily electricity in particular – was during the third quarter of the nineteenth century. This was the issue that really divided the likes of Lobb from the likes of the *Lancet*'s editors. Specialism appeared dangerous to the medical profession's social position because generalism was linked to gentility. The gentleman doctor was meant to be a rounded, cultured man of parts, whose practical skills were leavened with a good dose of liberal education. Specialism, on the other hand, looked narrow and pettifogging. This was particularly true of medical electricity which as far as many medical men were concerned seemed far too much like getting their hands dirty. It all came down to who the right kind of people were to apply electricity to the body – and where they should do it.

Looking at the different sorts of places that someone like Constance Phipps could come across electricity can therefore tell us something quite valuable about electricity's place in mid Victorian culture. It can tell us how electricity was experienced. By the 1860s electricity was starting to become a more familiar feature of the Victorian cultural landscape. It was starting to come out of the laboratory in a number of novel and far more visible and tangible guises. The next two chapters will survey the electrical scene that a young woman of the Victorian upper classes might have encountered: a world of exhibition and sensation. Electricity was something that thrilled and teased the senses. But it was also, as Constance well knew, a way of keeping the body in check. Doctors offered electricity as an antidote to wayward spirits. But whatever else it was – encountered as spectacular sensation or as a technology for discipline – electricity was intimately corporeal. It is, unfortunately, impossible to know exactly what Constance knew about electricity, except of course that it was meant to cure her ills. It is possible to make some educated guesses, nevertheless, based on who she was and the circles in which she moved.

ELEVEN

ELECTRIC FRONTIER

What would a girl like Constance Phipps in mid-Victorian England know about electricity? Even if knowledge of the specifics of what Constance knew is no longer possible, there is still quite a lot to learn about the expanding place of electricity in the world she lived in. If we do not know exactly what she knew, we can certainly examine what she ought to have known – what her surroundings, family and education made available to someone like her. By the 1860s the electric fluid was, if not precisely everywhere, then certainly leaking out of laboratories and workshops in more and more places. Electricity turned up in books and magazines, it appeared on stage at the theatre, it dazzled crowds at popular lectures and exhibitions. When Constance was born, the first experiments were being made in laying underwater cables for the new electromagnetic telegraph between Britain and France. When she was 6, the first attempt was being made to lay a telegraph cable across the Atlantic, and when she was 14, the Atlantic cable became a fact. If she visited the 1862 International Exhibition in London – which is more than likely that as almost everyone else of her class and background did – she would have seen spectacular new electrical technologies on show from Britain, France, Germany and the United States. In the 1860s and 1870s, electricity was the future.[1]

What did this electrical future look like at the beginning of the 1870s? One of the decade's bestselling novels gives us something of a clue. Edward Bulwer Lytton's *The Coming Race*, published in 1871, featured a young American mining engineer lost underground and stumbling across a mysterious subterranean race. The Vril-ya, as they called themselves, possessed mysterious powers. They could communicate telepathically, they could fly, they could control the weather and their environment and they could kill at a distance. The key to the Vril-ya's dominance of their underground world was their ability to control and manipulate a mysterious power called vril. Their control of this power was central to their existence as individuals and as a culture; it defined them and gave their race its name. In case his readers missed the analogy, Bulwer Lytton left no doubt as to just what the mysterious power really was. Vril was, to all intents and purposes, electricity:

except that it comprehends in its manifold branches other forces of nature, to which, in our scientific nomenclature, differing names are assigned, such as magnetism, galvanism, &c. These people consider that in vril they have arrived at the unity in natural energetic agencies, which has been conjectured by many philosophers above ground.[2]

Vril, in so many words, was what the Victorians could look forward to once they really had cracked the secret of electricity.

Bulwer Lytton, as even his many critics acknowledged, had a rare gift for recognising which way the wind was blowing. This was the secret of his success as a novelist   only his close friend and rival Charles Dickens outsold him during his own lifetime.[3] With *The Coming Race* Bulwer Lytton had clearly struck a chord. The novel went through eight editions in the first year and a half after publication. Vril itself became embedded in culture in a rather more unexpected way. In 1886, when the manufacturers of a new brand of beef extract were casting around for a name to define their product, they decided on Bovril (derived from bovine vril) as the ideal candidate. Constance Phipps was tragically denied the pleasure of Bovril on toast, having died a few years before it was first manufactured, but she was nevertheless exactly the kind of reader for whom *The Coming Race* was written. Bulwer Lytton depended for the book's verisimilitude on readers like Constance recognising vril as electricity, as well as understanding its very bodily resonances. For vril was not just electricity, it was 'mesmerism, electro-biology, odic force &c.' too, so that the Vril-ya could 'exercise influence over minds, and bodies animal and vegetable, to an extent not surpassed in the romances of our mystics'.[4] What Bulwer Lytton saw and his readers understood, surveying the electrical world around them, was that electricity and their own bodies were closely intertwined.

As far as Constance and her contemporaries were concerned, Victorian England could already perform with electricity many of the miracles that the Vril-ya had achieved. The electromagnetic telegraph had made instantaneous communication over large distances a reality – in the popular mind at least. In practice telegraph communication, whilst hugely faster than any previous method, was not quite immediate. Telegraph engineers spent much of their time coming up with new technical fixes to repair the telegraph signal's tendency to become unintelligible over long distances. The image was what mattered here though, with the telegraph seeming to make the world quite literally a smaller place.[5] It certainly looked as magical as vril might be to many of its enthusiasts. The Bishop of Llandaff rhapsodised as early as the 1840s about the way in which the telegraph revealed modern civilisation's superiority as compared both to the antics of medieval magicians, or the claims of contemporary oriental charlatans – far exceeding 'the feats of pretended magic and the wildest fictions of the East'.[6] As another pundit, the journalist and medical man Andrew Wynter, put it, the electromagnetic telegraph was 'a spirit like Ariel to carry our thought with the speed of thought to the uttermost ends of the earth'.[7] He was far from being the only commentator to regard the telegraph – and electricity in general – as simultaneously modern, magical and intimately corporeal.

The telegraph was also seen to be just like us. As Wynter put it, the Electric Telegraph Company's London headquarters were abstract intelligence embodied and personified. Visiting their offices jammed:

> between lofty houses at the bottom of a narrow court in Lothbury, we see before us a stuccoed wall, ornamented with an electric clock. Who would think that behind this narrow forehead lay the great brain – if we may so term it – of the nervous system of Britain, or that beneath the narrow pavement of the alley lies its spinal chord, composed of 224 fibres, which transmit intelligence as unperceived as does the medulla oblongata beneath the skin.[8]

The telegraph system would 'like the great nerves of the human body, unite in living sympathy all the far-scattered children of men'.[9] Victorian readers like Constance were encouraged to see their own bodies as the original templates from which electrical miracles might be drawn. As one American pundit put it, the telegraph, 'in its common form, communicating intelligence between distant places, performs the functions of the sensitive nerves of the human body'. But it could do more than that, it could act as the:

> motor function, or to produce effects of power at a distance; and this is also con- nected with the sensitive function, through a brain or central station, which is the reservoir of electric, or nervous power for the whole system. We have thus an excito-motory system, in which the intelligence and volition of the operator at the central station come in to connect sensitive and motor functions, as they would in the case of the individual.[10]

Bulwer Lytton would certainly have been familiar with the fascination that elec- tricity exerted on his readers. This was what made it such a compelling theme for his novel, after all. He would also have been aware, as Constance almost cer- tainly was, of the pull it exerted on some of his aristocratic and moneyed peers. One such peer, William George Armstrong, had made his fortune selling ships and guns to the navy and army. Knighted in 1859, he was eventually elevated as the first baron Armstrong for his services to British industry and military power. More importantly, he turned his estate at Cragside in Northumberland into an electric paradise. By 1871 he was using water from a reservoir on the estate to run a pow- erful Siemens dynamo, and was producing electricity for the house and grounds. Cragside was a showcase that put the electrical future on display. Armstrong, after all, had first made a name for himself as an electrical showman, putting his hydro- electric machine through its paces and producing shocks and sparks at the Royal

Polytechnic Institution in the 1840s. At first the electricity was used to power machinery on the estate, but by the end of the 1870s Cragside was illuminated by electric arc lights. Armstrong was not the only peer with a yen for electricity either. The Marquess of Salisbury, future Tory prime minister, was electrifying his own estate at Hatfield by the 1870s too.[11]

As early as the 1830s, electrical entrepreneurs had been speculating wildly about the possibilities of electrical travel. The first steamships had barely started plying their way across the Atlantic before electricity was being touted as a rival to steam. After all, 'half a barrel of vitriol, and a hogshead or two of water, would send a ship from New York to Liverpool; and no accident could possibly happen, beyond the breaking of the machinery, which is so simple that any damage could be repaired in half a day'.[12] There were plans to put electric locomotives on the railways too. In 1851 in the United States, the plausible federal patent officer Charles Grafton Page succeeded in persuading a usually parsimonious and risk averse Congress to part with $50,000 to finance the building of a line between Washington and Baltimore to test his electric train. At the Great Exhibition in Hyde Park's Crystal Palace in the same year, the Danish inventor Soren Hjorth exhibited an electromagnetic engine that moved the exhibition judges to exclaim that they could 'not help flattering ourselves that the attainment of this mysterious motive force will soon be followed by the making it available for practical purposes'.[13] In pride of place, overlooking the Crystal Palace's cavernous south transept was a huge electric clock, which not only kept its own time, but regulated the movements of several other timepieces scattered throughout the galleries.

Constance was born just a few years too late to enjoy the exhibits at the Crystal Palace – though she might well have enjoyed a visit to the glass cathedral after it was rebuilt in Sydenham a few years later. She is quite likely to have visited London's International Exhibition of 1862. Her father, being who he was, must surely have gone and bringing along the family is just what an improving Victorian paterfamilias would have done as well. If she was dragged around the exhibits (she would have been 9 at the time) she would have seen plenty of electricity. It was impossible to avoid electricity even just getting into the exhibition hall. The 'magnetic telltale' of Professor Wheatstone was attached to some of the turnstiles … each visitor, on passing through it, unconsciously and telegraphically announced his or her arrival to the financial officers in whose rooms were fixed the instruments for receiving and recording the liberated current'.[14] Constance may have been bored to tears by the extensive displays of telegraph cables, but might also have been amused at the prospect of sending her own telegraph messages from inside the exhibition, or at the dynamos producing electricity for electric lighting. Constance was certainly not at Philadelphia's Centennial Exhibition in 1876, but she would probably have read of the star exhibit – Alexander Graham Bell's ingenious device for electrically transmitting the human voice down a telegraph cable. It was lampooned in *Punch* and breathlessly reported in the journals.

The one electrical feature of the 1862 exhibition that seems to have provoked most public comment was the electric light. Reporting on the first of the shilling days at the beginning of June (a tradition continued from the 1851 exhibition), *The Times* remarked that the crowd surrounding the display of electric lights was so thick 'it was almost impossible to force a passage'.[15] A week later and the 'crowd was as thick as at the pit-door of Drury-lane on a Boxing-night, and at some periods of the day it was next to impossible to elbow one's way though it'.[16] In another week:

> the crowd round that part of the passage is so great there is hardly any possibility of getting into its proper focus, and arrangements are being made to place a reflector in the gallery at the western end of the nave between the two great organs. The machinery will remain and work in its present position, and the light will be transmitted along a wire to this point, whence it will be reflected so as to be visible all down the nave to the western end.[17]

It had to be moved from there in the end as well, since the 'intolerable brilliancy is literally blinding, and not only blinding at the time, but for some seconds after'.[18]

By the beginning of the 1860s, electric light was starting to displace gas lighting in its starring role at the annual London Illuminations. By 1863 electricity was used to light up the Monument. The authorities at St Paul's cathedral roped in Professor Pepper from the Royal Polytechnic Institution to install their electrical illuminations. Under his supervision, three 'of Groves' [sic] amalgamated zinc batteries, each containing 40 cells, were placed on the roof of the nave to light the western front of the dome, and a fourth was fixed on a platform which had been specially erected immediately under the exterior of the ball and cross'. It was not a complete success. The 'illumination on any particular spot was intermittent, and it was evident that many more batteries and lamps would have been required to produce the continued lighting up which the public had expected'.[19] *The Times* still applauded, however. It was the:

> electric light, as it had been perfected by repeated modifications of its chymical conditions, which showed off the noble proportions of St. Paul's and the Monument, and the view of the Cathedral, as seen by those who were wise enough to take their stand on one of the bridges, fully justified the experiment, though it was not entirely successful.[20]

The electric light had its detractors, though, particularly when shone upon living flesh rather than dead stone. *The Times* further acknowledged after one demonstration that its 'effect upon the human countenance was … by no means favourable … casting on all a strange unearthly hue'.[21] One of their correspondents was more forthright, complaining that the electric light 'makes benevolent men with rosy faces look like animated blue corpses of great malignity'.[22]

It is interesting that the dean and chapter of St Paul's cathedral turned to Professor Pepper and the Royal Polytechnic Institution for help in installing their almost, but not quite, spectacular display of electric light. They were not the first or the only ones. In the dull autumn days of October 1861 the Middle Temple organised an extravagant *conversazione* to celebrate the opening of their new library. The Prince of Wales attended but left immediately after dinner and before the real show of the evening began (presumably to find his own, less salubrious, entertainment). Who to turn to but Professor Pepper? Pepper duly laid on 'a great variety of scientific objects and works of art, which attracted great attention'. The display included a 'beautiful collection of specimens of diamonds, pearls, rubies, and other precious stones, illuminated by the electric light'. Outside, the 'Little Garden, with its rippling fountains, tasteful arcades, and masses of pompones, was beautifully illuminated by the electric light'.[23] More than a thousand guests were there to admire the extravaganza; Pepper was clearly making a name for himself as the man to whom people turned when they wanted to know about electricity – and the Polytechnic as the place to go for electrical entertainment and edification.

By the 1860s John Henry Pepper was firmly entrenched as London's most popular purveyor of spectacular science.[24] Constance may have been the wrong gender to get much out of his bestselling *Boy's Playbook of Science* but she must surely have attended some of his scientific shows.[25] He had started his career as one of the Polytechnic's cadre of lecturers in 1847, and by the 1850s he had been elevated to director – a position he maintained until he fell out with his fellow managers in 1872, decamping briefly to the rival Egyptian Hall before setting off on lecturing tours of North America and Australia, and finally settling in Brisbane in the more humble capacity of a consulting chemist. Back in the 1860s, whilst the dean of St Paul's and the gentlemen lawyers of the Middle Temple were courting his services, Pepper was at the pinnacle of his career. At the end of 1862 he had unveiled what was to become his best remembered showpiece – the optical illusion now known as Pepper's ghost. The illusion was simplicity itself. A plate-glass sheet was suspended at a 45-degree angle in front of the stage. An actor, hidden in the pit between stage and audience and lit up by the light from a powerful magic lantern, played the ghost. The actor's reflection appeared on the plate glass, making it look as if the ghost was onstage. By all accounts it was a highly convincing illusion; the 'effect of the first appearance of the apparition on my illustrious audience was startling in the extreme', Pepper later recalled.[26] Pepper used the ghost as the highlight of his *Strange Lecture* (the title deliberately evoking Bulwer Lytton's recently published ghost novel, *A Strange Story*), which ran the gamut of electrical and optical tricks to dazzle and confuse the audience.

The ghost would have been the highlight of any visit to the Polytechnic throughout the 1860s, though there was still plenty more to see. The Polytechnic was particularly known for its Christmas entertainments and its magic lantern pantomimes were very popular. The Polytechnic under Pepper's directorship offered its visitors a well worked out and reliable mix of new inventions and familiar favourites, along with carefully choreographed scientific lectures and magic lantern perform-ances. A typical programme of scientific entertainments during the 1860s or 1870s might commence with an 'inspection of Models, Cosmoramic views &c.', followed by a 'demonstration of new Inventions'. The revels might then continue with magic lantern lectures by J.L. King on the 'Progress of Royalty in India' or 'Astronomy' before drawing to a close with J.W. Harman's rendition of 'Gabriel Grubb and the Grim Goblin', complete with spectacular dissolving views and mechanical magic lantern effects. Visitors could try out their capacity to withstand electric shocks or examine the latest telegraphic apparatus. There were demonstrations of electric light, of course, as well as models of locomotives and boats. It was a popular after-noon's entertainment for children – or for their harassed governesses and tutors. *The Times* recommended it regularly each year, and more than one eminent late Victorian natural philosopher recalled in their memoirs being towed around the Polytechnic as a particular childhood treat.[27]

In April 1869 Pepper unveiled the most extravagant electrical spectacle yet when he put the Polytechnic's newly acquired Great Induction Coil through its paces before a select audience. It was a bravura performance and, suitably, was breathlessly reported in the press. Spectators 'oohed' and 'aahed' as Pepper worked the coil to generate indoor lightning: 'a spark, or rather a flash of lightning, 29 inches in length and apparently three-fourths of an inch in width, striking the disk terminal with a stunning shock.' This bolt of artificial lightning from the coil was so powerful that it could penetrate through 5in of solid plate glass. When the terminals were brought close together, 'the discharge appears to issue more slowly as a gush of waving flame, and this flame may be blown away in a broad sheet, leaving the actual line of discharge unaffected and visible by its different colour'. The new instrument was destined to 'be a source of endless delight and wonder, and enable Professor Pepper to display effects, beautiful or terrible, such as have never been seen before'. The apparatus and the whole performance surrounding it represented 'a triumph of skill and knowledge of which English science may be justly proud'.[28] This sounds like a typical Pepper performance: carefully calculated, contrived and rehearsed to impress his audience. Much of the press coverage was couched in identical language too; one suspects that Pepper had made quite sure that the invited correspondents were suitably primed.

The monster coil was hailed in the press as the future of electricity. It would 'not only amuse audiences, but will be diligently used at other times to promote the researches of electricians and physiologists'.[29] When in the hands of author, performer and medical reformer Benjamin Ward Richardson it seemed well set to achieve just that. Richardson borrowed the coil from Pepper and the Polytechnic for a series of spectacular experiments that dug deep into the connections between flesh and the electric fluid. Richardson was something of a medical jack of all trades: a prolific writer and commentator on affairs of the body and the body politic.

In 1876 his tract *Hygeia: The City of Health* described the ideal hygenic city, with an electric telegraph connecting hospitals with fire stations, factories, theatres and other points of danger. He would not have liked Harry Lobb's galvanic hospital; there was no room for specialist hospitals in the city of Hygeia.[30] In fact he had suffered bruising encounters with Lobb already, as Lobb challenged him for priority in using galvanism as an anaesthetic for tooth extraction. In a widely reported series of lectures at the Polytechnic in 1869, shortly after the Great Induction Coil had been inaugurated, Richardson set out to show just what the relationship between electricity and the body in life and death really was.

This was electrophysiology, Polytechnic style. Richardson set out to investigate the impact of lightning strikes on the body by manipulating the artificial lightning produced by the monster coil. He wanted to show his audience the effect of galvanism on different kinds of tissue – and he demonstrated the outcome to spectacular effect, making full use of the Polytechnic's substantial resources in the process:

> I place in glass tubes, a foot long and of equal diameter, portions of animal substance – blood, muscular fibre, brain matter, spinal cord, gelatine, water, fat; I arrange that the mass of each substance shall be the same. I pass a metal conductor the same distance into each, and I carefully insulate the tubes at both ends. I now make these tubes form part of the circuit of the coil and, acting on the very happy suggestion of Mr. Tobin, I interpose between the poles two of Gassiott's [sic] electric fountains or cascades. When the room is darkened, see how beautiful is the light as it streams over the glass within the globe; we are using at this moment a metallic conductor. See, now, the light is decreased, and the current from the coil, instead of making its way silently, flies across from a point to a point; we have interposed our tube containing fat, and the current, resisted by that, strikes across. See, again, the fountain is nearly as beautiful as at first; we have removed our tube containing fat, and interposed blood. See, again, the light is less; we have changed blood for distilled water. Lastly see the difference between blood and spinal cord.[31]

Gassiot's 'electric fountains' would have been familiar enough to Richardson's audience. The cascade, as the arrangement was more usually known, was another outcome of John Peter Gassiot's continued preoccupation with spectacular electrical displays that had been so evident in his electrical soirées during the 1840s. To achieve the effect, Gassiot lined a glass cup with tinfoil and placed it inside an air-pump with an electrode placed at its mouth. When the terminals of the induction coil were attached and the pump evacuated:

> at first a faint clear blue light appears to proceed from the lower part of the beaker to the plate; this gradually becomes brighter until by slow degrees it rises, increasing in brilliancy, until it arrives at that part which is opposite or in a line with the inner coating the whole being intensely illuminated. A discharge then commences

from the inside of the beaker to the plate of the pump in minute but diffused streams of blue light; continuing the exhaustion, at last a discharge takes place in the form of an undivided continuous stream, overlapping the vessel as if the electric fluid were itself a material body running over ... streams of lambent flame appear to pour down the sides of the plate, while a continuous discharge takes place from the inside coating.[32]

Henry Noad described Gassiot's Cascade experiment as 'one of the most beautiful that can be made with the Induction Coil'.[33] Electricity at the Polytechnic, or in Henry Noad's lectures, was all about bodily experience; it was about seeing and feeling.

Gassiot's cascade was one of the most spectacular in a line of novel electrical demonstrations that electrical showmen developed from the 1850s onwards. They were taking advantage of a new and more powerful version of the induction coil developed by the German electrician Heinrich Rühmkorff, and some of these displays used everyday bits and pieces to spectacular effect. The electrifying lecturer Henry Noad described how, if the discharge from a coil:

be made to pass over a lump of white sugar, or a crystal of alum, they will be beautifully illuminated; if through a fine iron two or three feet long, suspended by silk threads in a festoon, sparks, accompanied with brilliant scintillations, occur at every link. Should the chain be rusty, the brilliancy of the effect is increased.[34]

When a piece of phosphorus was placed in a vacuum and ignited with electricity from the coil:

the effect produced is that of a number of cones of light chasing each other from below upwards, and *vice versa*; sometimes they are flat tables of light, an inch or more apart; sometimes they are rings apparently revolving or oscillating and vanishing one into the other.

And if a little air was carefully introduced into the vacuum at the end of the experiment, 'a brilliant stream of magenta-coloured light will gradually blend with the whole'.[35]

These stunning demonstrations were the outcomes of experiments carried out by Gassiot, as well as other electricians, like the advocate of the correlation of forces, William Robert Grove, as they investigated the properties of electrified gases. Glowing gases had a long history as part of the electricians' armoury of technologies of display. In the eighteenth century, electrical showmen demonstrated an artificial aurora borealis captured inside a glass egg. Something like this, in all likelihood, was also behind Georg Bose's famous beatification effect in which the subject of his electrical demonstration appeared crowned with a glowing halo. The electricity from the powerful new generation of Rühmkorff coils developed during the 1850s did not just make gases glow as the current passed through them, captured in sealed glass tubes and cylinders, but revealed patterns of light and dark. Gassiot and Grove were both interested in understanding these mysterious striae and their origins, and in producing flamboyant displays like Gassiot's cascade at the same time. Others, like Warren de la Rue, Cromwell Varley and William Crookes, continued along the spectacular electric trail blazed by Grove and Gassiot. Varley was keen to experiment with ways of using photography to capture the continually shifting patterns of light from these electric discharges. Whilst Crookes thought that streams of glowing matter, torn apart by electric forces, were behind the effects. To prove his point, he made discharge tubes with a little glass railway suspended between the electrodes, along with a tiny miniature locomotive equipped with a paddle wheel. If he was right, he claimed, then pressure from the streams of radiant matter would push the little locomotive along the railway from one end of the tube to the other – and it did.

This world was the electric frontier as someone like Constance Phipps might have seen it. It is impossible to know, of course, just what she really did encounter; her adolescent experiences might have left her fascinated by electricity, or might have alienated her just as completely. But Constance did not even need to go to the Polytechnic Institution, or to visit the metropolis during the London Illuminations to be dazzled by electricity (though it is difficult to believe that she did not). She could read all about it instead and would have found literary electricity even more difficult to avoid. The variety of electrical literature was impressive and wide ranging. Bulwer Lytton's electrical blockbuster, *The Coming Race*, would have made little sense to its readership had they not already been familiar with galvanic matters from other sources. There was Pepper's *Boy's Playbook of Science*, squarely aimed at a juvenile male readership; perhaps those who had already been dragged through the Polytechnic on improving jaunts by governesses. Or, as in the case of the future Lord Rayleigh who would become James Clerk Maxwell's successor as Cavendish professor at Cambridge, by an elderly aunt.[36] Journals like *Macmillan's Magazine* that served up an eclectic mix of short stories, serials, poetry, politics and armchair tourism to their readers also included a good helping of electricity along the way. Notably, electricity was familiar, controversial and topical enough to be frequently satirised in *Punch*.

This was a world of tangible, earthy and corporeal electricity, leavened with a hint (or more) of theatrical greasepaint. It was not the only world of electricity in mid-Victorian Britain. In October 1871, James Clerk Maxwell delivered his inaugural lecture as the first Cavendish Professor of Experimental Physics at the University of Cambridge. Maxwell took advantage of the opportunity to look down his nose politely – though nonetheless firmly for that – at the culture of bodily electricity. There were two kinds of science, he said. On the one hand there were 'experiments of illustration', which aimed 'to present some phenomenon to the senses of the student'. On the other hand were those 'experiments of research ... those in which measurement of some kind is involved', which were 'the proper work of a physical laboratory'.[37] He made the same point again a couple of years later when anonymously reviewing *Electricity and Magnetism*, the new textbook by his friend, telegraph engineer and economist, Fleeming Jenkin:

> The author of this text-book tells us with great truth that at the present time there are two sciences of electricity – one that of the lecture-room and the popular treatise; the other that of the testing-office and the engineer's specification. The first deals with sparks and shocks which are seen and felt, the other with currents and resistances to be measured and calculated.[38]

As far as Jenkin and Maxwell were concerned, all people like Noad or Pepper did was produce 'apparently incoherent series of facts' and 'disjointed experiments' that were of no real or practical use to anyone.[39]

Maxwell was treading quite carefully in his inaugural lecture. He certainly did not want either potential students or university bigwigs to imagine that the newly endowed Cavendish laboratory would be devoted to producing a succession of Professor Peppers. But he did not want to go the whole hog with his friend Fleeming Jenkin and be seen turning the sons of gentlemen into hard-nosed, practical but dirty-handed telegraph men either. The Cavendish would have to plough a different furrow to make electricity respectable in the suspicious eyes of high-minded Cambridge dons. Maxwell needed to turn the Cavendish into a temple for a new kind of electricity, and a training ground for a new breed of electrical acolytes who would be divorced both from the grime of the engineer's factory and the showman's chicanery. It would not be easy, as Maxwell was himself the first to acknowledge:

> It is not without great effort that a science can pass out of one stage of its existence into another. To abandon one hypothesis in order to embrace another is comparatively easy, but to surrender our belief in a mysterious agent, making itself visible in brilliant experiments, and probably capable of accounting for whatever cannot be otherwise explained; and to accept the notion of electricity as a measurable commodity, which may be supplied at a potential of so many Volts at so much a Farad, is a transformation not to be effected without a pang.[40]

Maxwell and his successors at the Cavendish succeeded, nevertheless, in producing a new science of electricity that was both incorporeal and firmly grounded in disciplined bodies; both abstract and eminently practical.

The brave new electrical world that Maxwell was intent on creating at Cambridge was far removed from the one Constance and most of her contemporaries experienced, even if it was still made up from some of the same ingredients. To the throngs who flocked to the Royal Polytechnic Institution, thrilled at the electric light illu-minating St Paul's, or were dazzled at the 1862 exhibition, electricity was precisely about 'brilliant experiments'. And these experiments were usually intimately cor-poreal. Electricity worked for its mid-Victorian audiences by engaging their senses, not their intellects. It was something to see, to touch and to feel jolting and thrilling through their bodies. This was why the Vril-ya's relationship with vril in Bulwer Lytton's bestseller made so much sense to his readers; it was how electricity might well have seemed to people like Constance Phipps, too. Electricity was all about showmanship and performance as far as Pepper's audiences were concerned. It lit up the sky and disturbed the body's rhythms. It broke boundaries of time and space: the new Atlantic cable promised to throw a girdle round about the earth in forty minutes like some Shakespearean sprite. But if electricity was wonderful and magi-cal, it was a magic that could be controlled. That was what all the commentators said: the real wonder of electricity was that men could manage it. They could use it to domesticate their world and they could use it to discipline their bodies as well.

TWELVE

MACHINERY OF THE BODY

Even if we can do no more than guess about Constance Phipps' knowledge and experience of electricity more generally, we know for certain that she knew how electricity felt as it flowed through her body. She had, after all, been electrified by her father following Harry Lobb's instructions on the use of Pulvermacher's chain batteries. Constance was quite well placed, therefore, to appreciate and understand electricity's bodily impact. This sort of appreciation of electricity's peculiarly corporeal quality appears to have been quite widely shared. Electricity was something that one felt – if not for therapy, then for fun. A cartoon from the 1850s makes this point nicely. Entitled 'Christmas Holidays at the Polytechnic: The Electric Machine', it shows a man dressed in a sailor's uniform grasping the poles of an electrical machine, surrounded by a grinning festive crowd. The sailor hangs on grimly, a look of pop-eyed determination on his face. Standing next to him is another sailor, egging him on. It is easy to guess what is happening here. One sailor has challenged the other to take a charge of electricity and the other is doing his level best to win the bet. The scene is particularly telling, though, in its easy acceptance that this was how electricity was to be experienced. For a joke, for a bet – and for more serious purposes too – electricity was about the body.

In the body, or out, electricity in the 1860s could often be about keeping things under control as well. From the 1840s onwards, telegraphy had been touted by its promoters as an agent of social order. One of the telegraph's early success stories was its role in capturing a fugitive murderer, John Tawell, before he disappeared into London's murky underworld. He was seen getting on the train at Slough, where the crime had been committed, and a quick-witted telegraph operater sent the message down the line to Paddington so that he could be followed and arrested as he left the train. The telegraph that kept social undesirables in check was routinely imagined as a nervous system for the country, and it was exactly this capacity to keep a lid on dangerous tendencies and maintain control that made this nervous analogy such an appealing metaphor. This was, after all, how many Victorian physiologists thought about the nerves – they were the conduits for conveying the mind's instructions to an occasionally recalcitrant body. This was how electrotherapists like Thomas Laycock sold their electrical treatments too. Electricity could be depended upon to step in whenever the mind and the nerves failed to perform their appropriate function, maintaining bodily discipline and proper decorum.

Where did mid-Victorians find out about the variety of electrotherapeutic fare on offer? Importantly, the mid-Victorian period saw an explosion of print.[1] Books, journals, magazines, newspapers and pamphlets flourished in the wake of new mass printing technologies. Most books on electrotherapeutics seem quite clearly aimed at a professional readership. They surely reached a wider audience than that, but probably not to any great extent. At the top end of the market, nevertheless, the prestigious quarterlies like the *Quarterly Review* or the *Westminster Review* would carry reviews of the latest medical electrical works. The *Literary Gazette* or the *Athenaeum* carried such material too, as well as broader scientific news and correspondence that included discussions of electricity and the body. Medical almanacs aimed at the common reader discussed electrical therapies, as did the new generation of popular magazines aimed at aspiring middle-class households. At the bottom end, there were pamphlets aplenty extolling the virtues of one electrical therapy or another. Everywhere there were advertisements. Negotiating this sea of information could be as fraught for the practitioner as it could be for the potential patient. Harry Lobb's career came to an untimely end over exactly this issue of where and how to publicise his activities. But anyone who wanted to know about electrotherapy – who did it, where they might be found and just what they offered – had more than enough information to guide their enquiries.

By the 1860s, the German émigré Julius Althaus was generally recognised as one of London's foremost advocates of electrotherapeutics. Althaus had been born in the tiny German city of Detmold, capital of the diminutive principality of Lippe, his father a Lutheran minister. His medical education had him crisscrossing the German states from Bonn to Göttingen and Heidelberg, before he eventually received his MD from the University of Berlin in 1855. Before settling in London he spent some time in Paris studying with the great French neurologist, and pioneer of hypnotism as a therapeutic tool for hysteria, Jean-Martin Charcot. There in Paris he may have picked up his interest in medical electricity, since that was one of Charcot's enthusiasms too. In London, Althaus forged close links with Robert Bentley Todd, one of the metropolis' leading physicians and an influential figure at King's College Hospital. Under Todd's patronage Althaus started to make a name for himself at King's College Hospital as an exponent and practitioner of medical electricity. This was just the kind of break that an ambitious doctor needed in the cut-throat competitive world of Victorian medicine. A hospital place, even unpaid, and the helping hand of an influential backer were vital ingredients in the making of a successful career. They certainly helped give Althaus the social clout that he needed to establish his hospital for epilepsy and paralysis in Regent's Park in 1866. By which time, Althaus was already the author of influential treatises laying down the law about the proper practice of electrotherapeutics.[2]

Electrotherapeutics was a matter for professionals, Althaus insisted. It was not a panacea that could be administered willy-nilly by just anyone when nothing else seemed effective. It was a potential cure for specific diseases and conditions related to the nervous system. Moreover, particular forms of electricity – generated by different instruments – were effective in specific instances. Continuous electricity (by which Althaus meant the current generated from a constant battery) had 'a powerful influence on the electricity of the nerve', for example. It was therefore 'by no means improbable that in many cases, especially of functional nervous affections, it may act in restoring the systemic current of electricity to its proper condition'.[3] The faradic current, on the other hand (by which he meant an alternating current derived usually from an induction coil or from a hand-cranked electromagnetic machine) was 'capable of disturbing the molecular equilibrium of the motor nerves and muscles, so as to produce the state in which they are physiologically active'. This meant that it could be used 'to re-establish or to ameliorate the lost or impaired function of the motor nerves and muscles'.[4] Doses of electricity needed to be 'exactly measured to suit the different constitution, age, and sex of the patient'.[5] The grey matter of the brain could 'be appropriately likened to a galvanic battery in which an electric current is generated, and the white matter to telegraph-wires which conduct the current to any place where it may be required'.[6]

Like Althaus, Herbert Tibbits was an advocate of what was called the 'local application' of electricity. By this was meant its application to particular limbs or parts of the body rather than to the body as a whole. Tibbits was a follower of the French neurophysiologist and electrotherapist Guilleme Benjamin Duchenne, popularly known as Duchenne de Boulogne, whom he described as the 'father of electrotherapeutics'. It is tempting to suppose, since very little is known of Tibbits' background and training beyond the hints provided by him in his various books, that Tibbits may have trained with Duchenne at some earlier stage of his career. He dedicated the second edition of his *Handbook of Medical and Surgical Electricity* to the Frenchman, as well as translating Duchenne's *De l'Électrisation Localisée et de son Application à la Physiologie* into English. Duchenne was particularly known, then as now, for experiments in which he used electricity to manipulate nerve and muscle to reproduce the specific expressions associated with particular emotions on the faces of his experimental subjects.[7] Like his mentor, Tibbits also argued that electricity worked by acting on specific nerves and muscles. It could 'either stimulate or soothe both nerve and muscle, according to its variety and mode of application; it will frequently restore voluntary movement, it will relieve pain, heighten temperature, recall sensation, coagulate the blood, and dissolve or slowly cause the absorption of tumours'.[8]

It was with practitioners like Althaus and Tibbits that their contemporary, Constance Phipps' electrotherapist Harry Lobb, was in competition. Superficially, the three had much in common. They were each passionate and prolific advocates of electricity in medicine, they published widely and all three were associated with one of the new-fangled and contentious specialist hospitals that proliferated in London during the middle of the century. Lobb, of course, with his London Galvanic Hospital, and Althaus and Tibbits with positions at two of the new institutions devoted to the treatment of nervous diseases. Althaus and Lobb both had help in their early careers through the patronage of eminent and well-connected doctors. Tibbits and Lobb both seem to have had some connection with Duchenne de Boulogne. Furthermore, both would ultimately sail too close to the wind, and end their careers in losing battles with the medical establishment in the form of the Royal College of Surgeons. It is a fair assumption that had Lord Normanby called on either Althaus or Tibbits rather than Lobb to treat his invalid daughter (as he might well have done), the treatment offered would have been broadly the same. Lobb, too, tended to advocate localised treatment; that is certainly what he offered Constance Phipps. To all intents and purposes, electricity was being offered as a tonic. It was 'a powerful stimulant to the jaded nerves, increasing their irritability and lending tension to the centres'.[9]

One way in which Lobb's view of electrotherapy did stand out from the norm was in his insistence that it was a useful remedy for male sexual misbehaviour. He shared with others the more or less conventional view that female hysteria as 'an affection of the nervous system … frequently the result of excitement of the generative organs' was amenable to electrical treatment.[10] More unusual was his emphasis on electricity as a cure for the diseases emanating from the solitary vice. He described the symptoms that betrayed a young man's addiction to masturbation:

> he is highly nervous, and retiring to such a degree that it becomes noticeable by his friends and companions; he instinctively avoids the society of females, and feels a shame when introduced to them, involuntary flushings add to his discomfort, and he is overjoyed to escape from them.[11]

This was a social disease brought about by modern ways of living as much as by the moral turpitude of the sufferer. They often could not help themselves, living in 'large towns, where there are so many stimulants to sexual excitement' and as a result being 'kept in a continual state of desire'.[12] Electricity provided a means of reviving and revitalising the jaded parts before the disease spread from the organ to the brain, turning the masturbator into a drooling wreck.[13]

Lobb's morbid fascination with electricity as a corrective for youthful misdemeanours highlights the way in which electricity was so often regarded by its mid-Victorian promoters as a way of restoring discipline to otherwise wayward bodies. It was becoming apparent that bodies were not always quite – though clearly they ought to be – under the control of their owners. This was particularly true in the case of women, but Lobb's concerns make it clear that young men were not regarded as being entirely immune from the unfortunate condition either. Mid-Victorian physiology and social convention alike came together in seeing the body as something that needed to be reined in. It was no longer good enough to put loss of such control down to weak-mindedness either. A succession of early and mid-Victorian physiologists, like Marshall Hall, Thomas Laycock and William Benjamin Carpenter, had demonstrated irrevocably that the relationship between mind and body was too complicated for that simple approach, and that much of what the mind did to the body took place well beneath the surface of conscious thought and volition. This was just where electricity could be regarded as having a crucial role to play. Electricity operated beneath the surface as well, invisible except when it erupted through into the mundane world with spectacular effect. What electrotherapists offered from this point of view was a set of reliable techniques for manipulating the secret currents that coursed continuously between mind and body.

Just how electricity was related to the hidden mechanisms that held mind, brain and body together remained a tricky question. By the 1860s Britain no longer seemed as fragile in the face of possible revolution as it had done a few decades earlier. With the retreat of radicalism as an immediate political threat, materialism was not quite as dirty a word as it had been during the 1830s. This still did not make it respectable, nevertheless.[14] Simply equating electricity with the power of life remained a difficult tactic in polite circles. It did not either help that the slogan 'Electricity is Life' had by now been thoroughly hijacked by the electric belt brigade. Electrotherapists and physiologists needed to find a new way of talking about electricity and the body. One candidate was the stripped down language of correlation developed by William Robert Grove a couple of decades earlier. Grove, whose *Correlation of Physical Forces* went through two editions during the 1860s, had come up with *Correlation* as a way of talking about the relationship between nature's forces and the unity of nature without giving ground to the radicals.[15] Grove said nothing about the forces of vitality when *Correlation* was first published in 1846 (though he did in later editions). That did not prevent others from taking the hint and extending the correlation of physical forces from the realm of the inorganic to the organic.

First into the breach was William Benjamin Carpenter. In 1850 Carpenter was one of the rising stars of a new generation of ambitious physiologists who were champing at the bit to reform their moribund science. At this time he delivered a paper on the 'Mutual Relations of the Vital and Physical Forces' to the Royal Society and duly published it in the *Philosophical Transactions*. It was a paper that would probably not have seen print in *Transactions* only a few years earlier, had the Royal Society's victorious reformers, with Grove in the vanguard, not swept the old ruling clique and the long-standing secretary Peter Mark Roget away. Roget's malign and reactionary influence had been responsible, amongst other things, for keeping Marshall Hall's revolutionary views on the reflex action of the nervous system from sullying the pages of the Royal Society's august journal for much of the 1830s and 1840s. Carpenter argued that the language of correlation made it possible to course a chart for physiology that avoided the pitfalls of both material-

1 The Chemical Lecture Room at the old Botanical Gardens in Cambridge. This is where Thomas Weems' body was taken for galvanic experimentation following his hanging on Castle Hill. Courtesy of the Syndics of Cambridge University Library.

2 An example of electrical beatification. Benjamin Rackstrow, *Miscellaneous Observations, together with a Collection of Experiments on Electricity* (London, 1748).

3 Some of Luigi Galvani's experiments on frogs, demonstrating animal electricity, or galvanism. Luigi Galvani, *De Viribus Electricitatis in Motu Musculari Commentarius* (Bologna, 1792).

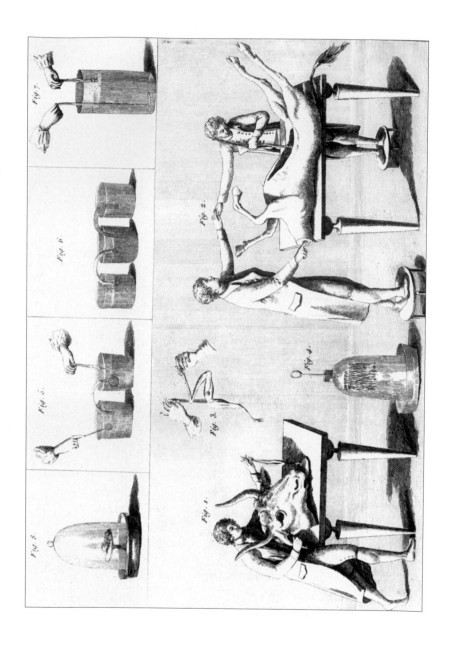

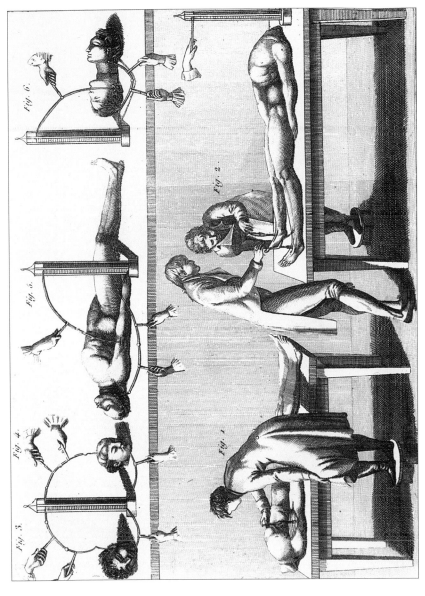

4 & 5 Giovanni Aldini's galvanic experiments on human and animal bodies. Note the use of frogs' legs as instruments to indicate the presence of animal electricity. Giovanni Aldini, *Essai Theoretique et Expérimental sur le Galvanisme* (Paris, 1804).

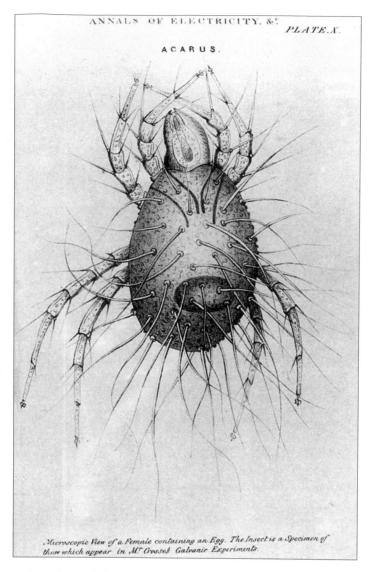

ANNALS OF ELECTRICITY. &c

PLATE.X.

ACARUS.

Microscopic View of a Female containing an Egg. The Insect is a Specimen of those which appear in M.r Crosse's Galvanic Experiments.

6 A specimen of *Acarus Crossii*, the insect that appeared during Andrew Crosse's galvanic experiments in 1836. *Annals of Electricity*, 1837.

7 The frontispiece of Henry M. Noad, *Lectures on Electricity* (London, 1844). A wide range of electrical apparatus is visible, including an impressive Armstrong hydro-electric machine in the foreground. Some electric eels are swimming in a basin on the table to the right of the picture.

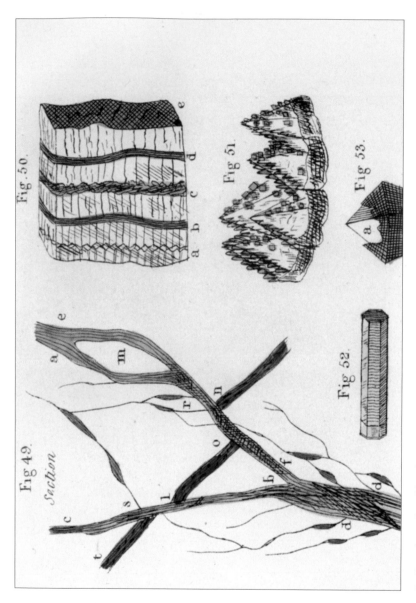

8 Crystals produced by electricity and representations of currents of electricity flowing through mineral veins in the earth. *Annals of Electricity*, 1838.

9 Michael Faraday lecturing at the Royal Institution The Prince Consort and the Prince of Wales are in the front row immediately facing him. *Illustrated London News*, 1854.

10 The frontispiece of
J.O.N. Rutter, *Human
Electricity* (London,
1854), demonstrating his
galvanoscope.

11 The great hall of
the Royal Polytechnic
Institution. Westminster
University Archive.

12 'Christmas Holidays at the Polytechnic'. Visitors to the Royal Polytechnic Institution enjoying themselves with the electrical machine. *Illustrated London News*, 1858.

13 The Giant Induction Coil at the Royal Polytechnic Institution, presumably with Professor Pepper lecturing. *Illustrated London News*, 1869.

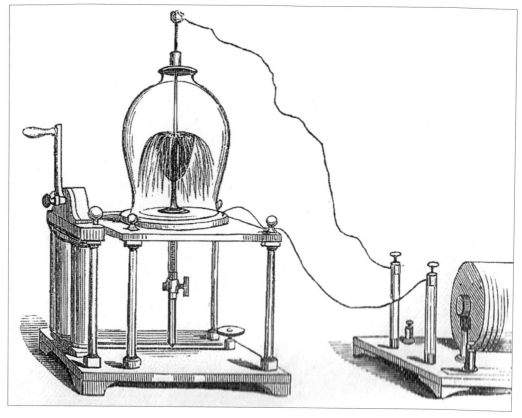

14 The Gassiot Cascade demonstrated in Pepper's *Boy's Playbook of Science* (1866).

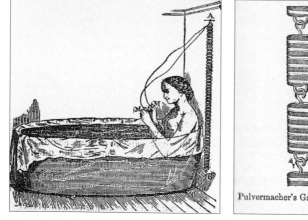

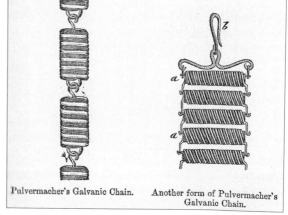

Pulvermacher's Galvanic Chain.   Another form of Pulvermacher's
                                   Galvanic Chain.

15 A demonstration of the Pulvermacher chain being used in an electric bath. *Practical Guide for the Electro-Medical Treatment of Rheumatic and Nervous Diseases, Gout, etc., by means of Pulvermacher's Hydro-electric Chains* (1856).

16 An advertisement for Cornelius Bennet Harness' electropathic belts and corsets, showing the Medical Battery Company's plush Oxford Street premises. *Illustrated London News*, 1891.

17 Medical apparatus combining galvanic and gymnastic therapy at the Institute of Medical Electricity. *Electrician*, 1889.

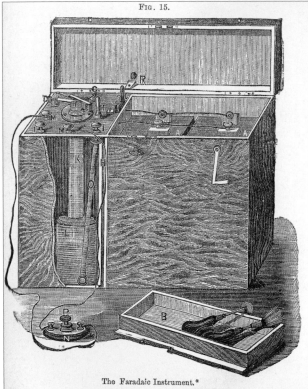

FIG. 15.

The Faradaic Instrument.*

---

* A. Cells shown by the removal of the compartment, B, for conductors and accessories.

18 A travelling doctor's electrical toolkit. Schall, *Illustrated Price List of Electro-medical Apparatus* (1896).

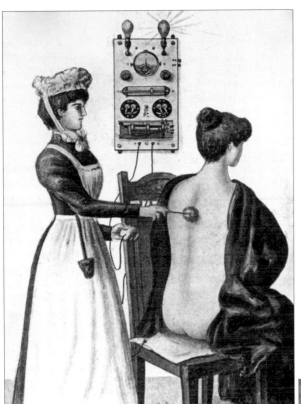

19 A nurse administering electrical treatment to a patient. H. Newman Lawrence & Arthur Harries, *Manual of Practical Electro-therapeutics* (1891).

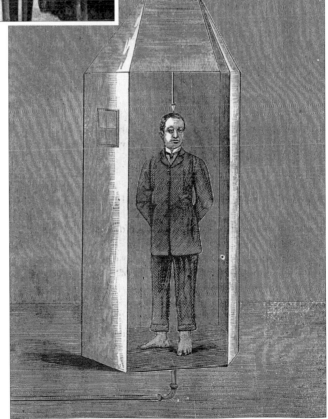

20 A plan for electrical execution. J. Mount Bleyer, *The Best Method of Executing Criminals* (1888).

21 The French physicist Arsene D'Arsonval demonstrating the use of the human body as a conduit for the transmission of electric power. *Electrical Review*, 1894.

ism and vitalism. Carpenter argued that all the various forces involved in organic process were modifications of a single vital force, and that this vital force was itself correlated to the forces operating on inorganic matter like electricity, heat, light and so on. Each of these forces was as closely correlated to the vital force as they were to each other. In fact, 'the chief distinction between their respective operations' was 'established by the speciality of the instruments through which they manifest themselves'.[16] So the forces that operated on the human body were similar yet different to the forces that operated on brute matter, in just the same way that those forces were similar yet different to each other.

   This was a clever manoeuvre. If it worked it meant that Carpenter could talk about relationships between vital and physical forces without being accused of confusing the one with the other, any more than someone using the language of correlation might be accused of confusing light and electricity. Harry Lobb used the language of correlation in his first publications on medical electricity. Correlation, pronounced Lobb, was 'a great step in the simplification of the study of science'. Electricity was 'convertible, under certain conditions, into every other force'. As a result, if:

> we want a stimulus to organic force, what so powerful as electricity? Will it not stimulate the growth of plants? Can we not cause the muscles of the dead to contract; cause the nerves to perform their function after death, even to digestion; lend health to the diseased, and life to the dying? Thus, then, we have in electricity a grand and universal panacea.[17]

In slightly more measured tones, Joseph Snape accounted for electricity's usefulness as an anaesthetic by citing Grove's view 'that what is commonly called the nervous fluid, correlates with the physical forces'.[18] Benjamin Ward Richardson invoked Grove's doctrine in his own accounts of the operations of electricity on the nervous system – and dedicated a volume of his *Asclepiad* to Grove in recognition of his debt.

   Hand in glove with this account of the relationship between electricity and the vital forces came the image of the nervous system as a telegraph. Whilst the telegraph's apologists turned to the nervous system as a convenient model to explain to the public how telegraphy worked, physiologists and electrotherapists turned the metaphor on its head for their own purposes. Grove himself, addressing a crowd of newly graduated medics in 1869, speculated that the galvanic battery was 'the nearest approach man has made to an experimental organism'.[19] This was perhaps no surprise, coming from someone who, despite being one of the grand old men of natural philosophy by the 1860s, doubtless remembered that they first rode to fame on the back of an improved battery. Harry Lobb suggested that the 'system of nerves and its ganglions may be likened to an electro-magnetic apparatus: the tissue change gives rise to the fluid, as in the battery: this is transmitted to the ganglion, where it is intensified, as in the magnetic arrangement'. He thought the similarity stopped there though, since 'the method of direction is peculiar to the ganglion'.[20] Lobb clearly thought enough of the metaphor to present a paper to the *British & Foreign Medico-chirurgical Review* a few years later, suggesting that the 'great impetus which the introduction of Ohm's formulae has given to the science of electro-telegraphy has encouraged me to make a like effort for medical electricity' and attempting to lay down practical electrotherapeutic rules by comparison with telegraphy.[21]

By the 1860s, however, talk of correlation looked a little antiquarian as far as many up and coming natural philosophers were concerned. The focus of natural philosophy had shifted from force to a new quantity – energy. Devotees of this new physics were dismissive of the vocabulary of correlation. Peter Guthrie Tait, one of the prime movers of the new physics and co-author with William Thomson of its bible, the *Principles of Natural Philosophy*, thought that Grove's work was 'humbug' and that his language was 'woefully loose and unscientific'.[22] James Clerk Maxwell even presumed to tick off Michael Faraday on his use of old-fashioned language: 'energy is the power a thing has of doing work,' he said, whilst 'force is the tendency of a body to pass from one place to another'. It was 'important to keep our words for distinct things more distinct', he insisted.[23] Despite this it seems that outside the relatively small group of mathematical physicists who developed the new science of energy, the language of correlation of forces seems to have been used interchangeably with the conservation of energy for much of the rest of the century. This was particularly the case with physiologists and others who were interested in exploring the relationship between physical and nervous forces and the powers of the mind.

Nevertheless, the language of the conservation of energy was used by its proponents to make sense of living matter. This was how Balfour Stewart, the first Professor of Natural Philosophy at the new Owens College in industrial Manchester put it:

> let us suppose that a war is being carried on by a vast army, at the head of which there is a very great commander. Now, this commander knows too well to expose his person; in truth, he is never seen by any of his subordinates. He remains at work in a well-guarded room, from which telegraphic wires lead to the headquarters of the various divisions. He can thus, by means of these wires, transmit his orders to the generals of the divisions, and, by the same means receive back information as to the condition of each.

Life worked like that too, insisted Stewart. Life was 'not a bully, who swaggers out into the open universe, upsetting the laws of energy in all direction, but rather a consummate strategist, who, sitting in his secret chamber, before his wires, directs the movements of a great army'.[24] All the forces at play in the human body were purely physical ones. Bodies were 'delicately-constructed machines' in which

transformations of energy were continually taking place. Life was always just out
of sight, though; it was the ghost inside the machine.

According to its proponents, there were still important lessons for society to learn
from contemplation of the conservation of energy. These lessons were particularly
pertinent to people like Constance Phipps. It was an essential principle of the conservation of energy that there was only so much of it around. There was only so
much energy in the universe and there was only so much energy in any single
individual system – like the human (and particularly the female) body. If women
squandered their limited sources of energy on frivolities like education and emancipation there would not be enough left for their proper function of childbirth and
nurturing; the race would suffer. As far as the mental physiologist Henry Maudsley
was concerned, this meant that a woman's place was ordered by nature, not society.
If 'it were not that woman's organisation and functions found their fitting home in
a position different from, if not subordinate to, that of men, she would not so long
have kept that position', he argued. This was:

> not a mere question of larger or smaller muscles, but of the energy and power of
> endurance of the nerve-force which drives the intellectual and muscular machin-
> ery; not a question of two bodies and minds that are in equal physical conditions,
> but of one body and mind capable of sustained and regular hard labour, and of
> another body and mind which for one quarter of each month during the best years
> of life is more or less sick and unfit for hard work.[25]

The iron law of the conservation of energy had built into its very fabric the intellectual and social hierarchy of the sexes. This was the kind of model that underlay
much electrotherapy too: medical electricity worked by replenishing nervous
energy that had otherwise become lost.[26]

Balfour Stewart – like most of the men who pioneered the new science of
energy in the British isles – was far removed from the kind of militant materialism
that had once been associated with talk about electricity and the body. Stewart, for
one, clearly thought that is was possible, and indeed necessary, to preserve an inef-
fable essence of life alongside its physical machinery. That was the whole point of
his telegraphic analogy. To many of its proponents, the conservation of energy was
a theological as much as a physical principle – energy was conserved because God
had created it and, therefore, only God could destroy it. This was a sober, Protestant
science.[27] The spectre of materialism was never far away, though, and returned to
centre stage in the summer of 1874 at the annual gathering of the British Association
for the Advancement of Science, held that year in devoutly Presbyterian Belfast.
John Tyndall, the saintly Faraday's successor at the Royal Institution and himself
an Orange-tinged Irishman from County Carlow in Leinster, delivered a presidential address that left the city's church elders gasping. Tyndall not only mounted
a trenchant defence of Darwin's doctrine of evolution by natural selection, but

insisted that physical evolution applied to the human mind as well. 'We can trace the development of a nervous system, and correlate with it the parallel phenomena of sensation and thought,' he insisted; we then 'see with undoubting certainty that they go hand in hand'.[28]

Tyndall's measured outburst was not the only thing to raise evangelical hackles that summer in Belfast. Thomas Henry Huxley – caricatured as Darwin's bulldog after his bruising encounter with Soapy Sam Wilberforce at the Oxford meeting of the BAAS in 1860 – delivered a sermon on automatism that simply denied the pious claim of Balfour Stewart, and his like, that there had to be something else going on besides the mere machinery of the body. This plainly was not so, said Huxley. Brute animals were merely 'conscious automata'. Any consciousness they possessed was 'the consequence of molecular motion of the brain' and 'an indirect product of material changes'. The soul – if brute animals had one – 'stands related to the body as the bell of a clock to the works, and consciousness answers to the sound which the bell gives out when it is struck'.[29] So far, so Cartesian, but Huxley's punchline left Descartes' philosophy standing. He could see no reason, he said, why the argument as applied to animals did not apply just as forcefully to men. In that case, 'all states of consciousness in us, as in them, are immediately caused by molecular changes of the brain-substance'. We were all:

> conscious automata, endowed with free will, in the only intelligible sense of that much-abused term – inasmuch as in many respects we are able to do as we like – but none the less parts of the great series of causes and effects which, in unbroken continuity, composes that which is, and has been, and shall be – the sum of existence.

So what if this philosophy led to atheism, challenged Huxley. What mattered was that it was true, in which case any persons who challenged it were simply liars 'whose opinions therefore were unworthy of the smallest attention'.[30]

Huxley's address was a step too far, for many of the previous generation at least. William Benjamin Carpenter, whose own youthful researches on the automatic nervous system were used by Huxley to bolster his arguments, thought that Darwin's bulldog was deluding himself. If Huxley were to be believed, Carpenter scoffed to the Metaphysical Society, then:

> the members of the British Association did not come together on that particular evening to hear Professor Huxley lecture, since no desire or will on their parts had anything to do with putting their bodies in motion to ride or walk to Ulster Hall; but that they were moved to do so by certain states of brain, which had no dependence on what they had previously heard or known of Professor Huxley, or on their expectation of an intellectual treat in listening to him, but was simply a consequence of molecular action excited through the retina of each by the image of the placard announcing the lecture, or through his tymapnum by a vocal communication.[31]

Talking to the members of the Glasgow Science Lectures Association a few months later, Carpenter was adamant that the body, machine as it was, was the instrument of the ego, separate from itself. The bodily machine was what mediated between the ego and the external world.[32]

Of course Carpenter did not deny that a great deal of what the nervous system did happened independently of the ego – that would have meant repudiating much of the work that had established his own reputation as a physiologist. Nevertheless, he insisted that there was a continuum between entirely automatic acts and volitional ones. He illustrated this by pointing to acts that had clearly once been volitional but now took place automatically. His friend the philosopher John Stuart Mill was a walking example of the phenomenon. 'I have seen John S. Mill making his was along Cheapside at its fullest afternoon tide,' he averred to the readers of the *Contemporary Review*:

> threading his way among the foot-passangers with which its narrow pavement was crowded, and neither jostling his fellows nor coming into collition with lamp-posts; and have been assured by him that his mind was then continuously engaged upon his System of Logic (most of which was thought out in his daily walks between the India office and his residence at Kensington), and that he had so little consciousness of what he was taking place around him, as not to recognise his dearest friends among the people he met, until his attention had been recalled to their presence.[33]

Neither Huxley nor his acolytes knew enough about the properties of matter to deny the existence of a conscious, independent will, Carpenter argued, pointing intriguingly enough to William Crookes' latest experiments on mysterious electrical rays as evidence that there were more things in Heaven and Earth than were dreamt of in Huxley's philosophy.[34]

Electricity – and the telegraph in particular – provided Carpenter with a fund of material that helped him explain the scope and the limitations of the electrical body machine. Nerves were structurally identical to telegraph cables. The 'cylinders of fatty matter' that surrounded them were:

> insulators, like the gutta-percha or other insulating material around our telegraph wires. Each of the telegraph cords that cross our streets, holds about sixty wires; and every wire has its separate origin and termination. As with these wires, so with the nerve-fibres; each has its separate origin and termination, and must therefore be insulated, in order that it may be functionally independent of those with which it is bound up in the same nerve-trunk.[35]

The electric link went beyond that, too. It was a common experience for telegraph clerks, according to Carpenter, 'when they have been for some time in the service, to work the instruments without conscious thought of what they are doing'.

They had become extensions of the telegraph apparatus, they 'read the words ...
pass them through their minds, and transfer them to the sending part of the apparatus, just as unconsciously and automatically as Wheatstone's transmitter does'.[36]
Carpenter was an admirer of the experimental physiologist David Ferrier, whose
researches at the West Riding Lunatic Asylum used electricity to stimulate different
areas of the cerebrum, exciting involuntary movement and pain in his experimental
subjects. The experiments showed how electricity could stimulate movements and
actions that would otherwise require the active volition of the subject, bolstering
Carpenter's view that the 'Cerebrum furnishes the "mechanism" of Thought and
Feeling; but its reflex action is directed and controlled by the Will, in proportion as
the Volitional power is developed'.[37]

Ferrier conducted his experiments on animals, rather than the human inmates,
at the West Riding Lunatic Asylum, side by side with efforts to use electricity as
a treatment for the insane. One of the reasons that Ferrier was carrying out his
experiments at the asylum was to try to identify particular parts of the brain with
specific mental disorders. There was an evolutionary story here too. The automatic
centres of the brain were older, more primitive, than those parts dominated by
the will. The primal machine was buried deep under layers of civilisation and self-
discipline, but it was still there and it took relatively little for the beast inside us all
to reassert itself, as Robert Louis Stevenson's *Strange Case of Dr Jekyll and Mr Hyde*
suggested. Electricity provided a tool both for understanding that primitive, bes-
tial machine and for containing it. Even as civilised and cultivated a mind as John
Stewart Mill's could relinquish control to the automatic system lurking deep inside
his brain. People like Carpenter worried that this was a modern disease. Both the
stresses and the blessings of industrial, urban culture might lead to degeneration
as minds and bodies either cracked under the strain or withered into uselessness
without the challenge of continuous competition. Insanity and 'nervousness' were
on the rise as the beast erupted to the surface. Electricity promised a way of putting
it back in its cage.[38]

This was electricity as Constance Phipps encountered it: as visceral and intimate.
In just the same way that the telegraph promised new ways of subjecting society to
surveillance, providing new tools for maintaining social order, electricity governed
the body. The image worked both ways for many Victorians precisely because they
associated discipline with oversight.[39] Just as the telegraph made possible the circulation of information through society, the nerves passed information back and
forth between body and mind. Men like Althaus, Tibbits and Lobb tried to present
themselves as the overseers of this bodily electrical exchange. Bodies needed this
kind of externally imposed discipline because people could not be trusted to keep
themselves under control. The shape of society offered too many temptations for
women and men alike. Constance would have known that as a young woman she
was particularly prone to losing control; she was at the mercy of her bodily desires
and, as a woman, was likely to lack the capacity to subjugate her carnality as men

did. It is important to remember that, contrary to popular prejudice, Victorians did not deny women's sexuality. Far from it – in some ways women were deemed to be more intensely sexual beings than men. It was precisely because of this that they were particularly in need of the bodily discipline that electricity, in the right hands, could offer.

Electricity was about more than discipline, however; it offered tomorrow to the Victorians. Playing with electrical fire allowed people like Constance to see the future. Whether or not Constance Phipps ever read Bulwer Lytton's *The Coming Race*, she would have found it quite difficult to avoid knowing something about the way it pictured the galvanic future. Bulwer Lytton's account consisted, after all, of electrical elements that would be pretty familiar to any reasonably well-read member of the middle and upper classes. The picture it offered would make perfect sense to someone who had visited the Royal Polytechnic Institution and seen Professor Pepper put the 'monster coil' through its paces, creating artificial lightning on demand, or had watched St Paul's cathedral shining in the electric light. Constance lived in a world in which electricity offered tantalising glimpses of the furthest edges of reality. Electricity made for a heady mix of metaphysics, technology and showmanship. It was everywhere; not just in laboratories like the Cavendish at Cambridge or Faraday's old haunts at the Royal Institution, but on hospital wards and in private consulting rooms, in factories and workshops, in theatres and exhibition halls. It was even starting to sparkle in aristocratic households. In Constance's case, we know electricity was in her bedroom.

Constance Phipps died on Thursday 31 October 1883, at Mulgrave Cottage on the family estates near Whitby, in Yorkshire. She was just 31 years old. It is telling of just how ephemeral she was, and how little her brief existence meant to the world beyond her family that the terse obituary notice in *The Times* even managed to get her pedigree wrong, describing her as the Marquis of Northampton's daughter. Constance never married, and when she died her parents were in Australia, the other side of the world from Whitby.[10] We do not know who was at her bedside. Her father had been made governor of the state of Victoria in 1879, her mother died there the following year and her father returned to England – he never accepted another public position, dying six years later in 1890. Despite the little we know of her and her apparent lack of consequence, Constance matters because her life provides a little window into the mid-Victorian electrical world and its peculiarly corporeal quality. After all, whatever else Constance might have known, seen or felt of electricity, whether she ever visited the Royal Polytechnic and saw the electric light or read earnest articles in magazines about the electrical future, we know that in the end her primary experience of electricity would have been utterly and intimately corporeal. It would have been the experience of a sick and frightened adolescent girl.

PART IV

MR JEFFERY

THIRTEEN

INTRODUCING MR JEFFERY

Sometime in 1892 a presumably rather nervous bank clerk, Mr D. Jeffery, wandered into the plush premises of the Medical Battery Company on Oxford Street and diffidently enquired whether it might be possible for him to have a consultation with a qualified practitioner. That is almost all that we know about Jeffery, not even his full name, except that he later stated that he had worked for his employers, the Union Bank, for twenty-nine years, so he must have been at least in his forties. He had seen one of the Medical Battery Company's ubiquitous advertisements extolling the virtues of its range of electric belts and other health-giving paraphernalia and wondered if a belt might help in his own case. As we have seen already, there had been a thriving market for electrical medicine throughout the nineteenth century. The surgeon Charles Wilkinson, who helped Giovanni Aldini with his electrical experiments on the dead, also advertised his own services as a medical electrician. At the risk of causing serious damage to their own gentlemanly and professional reputation, this is something no respectable medical man would have dreamed of doing a couple of generations later. Indeed by the 1840s, the editor of the *Electrical Magazine* was looking down his nose at the deluge of galvanic rings and belts that were beginning to flood the market. By the middle and latter parts of Victoria's reign, newspapers and magazines were full of advertisements extolling the virtues of a huge range of electrical contraptions designed to make their owners' bodies fit and healthy. Electrical health was a big, and lucrative, business.

In this crowded marketplace for electrical nostrums, the Medical Battery Company ruled the roost. It was certainly almost impossible to open a newspaper or magazine during the 1880s and 1890s without coming across one of their advertisements. They seemed to be everywhere. In fact, it is clear that the company had a cosy relationship with a number of publications. No wonder, therefore, that it was into their showrooms at no 52, on the corner of Oxford Street with Rathbone Place, that poor Jeffery wandered. By 1892 this was a company that really had succeeded in making the market its own. Jeffery explained at the desk that he was worried that he was suffering from what he later described as a 'rupture'. Having asked for a consultation with a qualified practitioner, he was introduced to a Mr Simmons. Simmons duly examined him, assured him that he did not have a 'rupture' after all, but persuaded him to purchase a belt as an insurance against any future problems. Jeffery allowed himself to be talked into the purchase and handed over £2, as well as an IOU for a further £3 5s. Some time later, unconvinced that the electric belt was in fact doing him any good at all, Jeffery consulted a doctor, who informed him that he was actually suffering from a hernia and that the electric belt he was wearing was probably doing him more harm than good. Angry at being deceived and misled by the 'expert consultant' Simmons, Jeffery refused to pay the outstanding IOU. The Medical Battery Company threatened legal action and in the face of Jeffery's continued refusal to cough up, took him to court to recover the unpaid bill. It turned out to be the biggest miscalculation in the company's history.

Jeffery was certainly not the first of the Medical Battery Company's customers to conclude that he had been short changed. Five years earlier, a correspondent styling himself 'a poor victim' wrote to the *Telegraphic Journal & Electrical Review* to air his grievances. The 'poor victim' described how upon seeing:

> a 'Wonderful' Electrical Belt extensively advertised by a manufacturer who styles himself an 'Eminent Consulting Medical Electrician', and who occasionally takes a whole page of *The Telegraph*, *Standard*, or *Daily News*, for the purpose of making known to suffering humanity the many virtues of his marvellous discovery, I must confess that I did not believe such astonishing statements were wholly unreliable.

A postal order for 3 guineas was duly sent, and in return the complainant found himself the owner of 'not an electric belt, but a thing which I should consider dear at five shillings'. An electrically minded acquaintance confirmed that he 'might as well have pitched the money into the gutter' for all the good the belt would do him. Even if 'joined up with the most sensitive testing instrument known to the profession', it 'would not show a trace of electricity'.[1] By the end of the 1880s the electrical and medical professional press were receiving a steady stream of correspondence complaining about the Medical Battery Company's sharp practices, loud in their indignation at this misuse of electricity. They seemed powerless to do much, however – and some correspondents even complained that their copies of the *Electrical Review* arrived with advertisements for the Medical Battery Company's products enclosed.

The 'Eminent Consulting Medical Electrician' behind the Medical Battery Company was a gentleman by the name of Cornelius Bennett Harness.[2] Relatively little is known about Harness' background, and what little there is must be taken with a pinch of salt since most of it emanates from those who led the virulent newspaper campaign that exposed him as a charlatan, and eventually led to his downfall in the aftermath of the Jeffery scandal. If these sources are to be believed, Harness had started his career as a salesman for a fancy goods company in London's Cheapside. He moved on from there to become a jeweller in Aldersgate. He was then employed for some time by the Pall Mall Electric Association – another seller of medical electrical paraphernalia – before setting himself up independently as the Medical Electrical Company, and taking over the Pall Mall Electric Association's assets.[3] According to the *Pall Mall Gazette*, who were at the head of the field in the campaign to bring his dubious business activities to light, Harness had been less than entirely straightforward in his dealings with his former employers. According to their story, the Medical Battery Company had originally been established as a blind. Its only purpose was to provide bogus competition for the Pall Mall Electric Association. Once the Medical Battery Company had become a going concern, however, Harness turned on his bosses, threatening to break the connection between the two companies and become a genuine rival unless the Pall Mall Electric Association's proprietors were willing to sell him their shares in the company.

Regardless of what really went on behind the scenes in shady backroom deals, we can at least be certain that Harness first established the Medical Battery Company in August 1881. He purchased the Pall Mall Electric Association from George Scott in exchange for £13,000 in shares and £481 in cash, and became the company's managing director. In 1884 he established premises on his own account on Regent Street selling electrical medical appliances. The Medical Battery Company was originally incorporated in 1885, joining together the Pall Mall Electric Association's premises in Holborn Viaduct with Harness' own business at 205 Regent Street; the amalgamated company moving to new premises in Oxford Street. The new company declared its value as £100,000, divided into 20,000 shares of £5 each, of which Harness owned £60,000. According to the articles of association, Harness was appointed managing director at a salary of £1,000 per annum. In 1889 the company was wound up voluntarily and reincorporated. Apparently, the reason for this was that Harness now wished to be able to treat patients on site and the original articles of association did not allow for this extension of the business. The new Medical Battery Company, according to its new articles of association, was committed:

> to provide for and carry out the treatment of diseases by means of galvanic, electric, and magnetic appliances, batteries, baths, medical and other apparatus, and by means of massage, hydrotherapic inhalation, mechanical exercise (Dr. Zander's or any other system), and other processes, mechanical or otherwise.[4]

By the late 1880s and early 1890s, if the prevalence of his advertisements and the constant and increasing attacks against him in the electrical press are any guide, Harness dominated the British market in electrical therapeutic appliances of all kinds. He was clearly a consummate publicist of his wares and of himself, losing no opportunity to emphasise his role as a great philanthropist devoting his inventive talents to the relief of mankind's suffering.

There could hardly have been a better demonstration of the Medical Battery Company's prestige and commercial success than the impressive premises they inhabited on Oxford Street. The building's imposing facade, prominently displayed in some of their advertising, was perfectly calculated to reassure potential customers of the establishment's respectability. This was bolstered as well by Harness' acquisition of the London Zander Institute and Zander's English patents for gymnastic appliances. Gustav Zander was a Swedish devotee of mechanical exercise as a means of maintaining and regaining health. He had developed an extensive routine of gymnastic exercise as well as inventing a number of exercise machines, and establishing institutes throughout Europe to promulgate his methods.[5] His was a name to be reckoned with. Harness promptly renamed his establishment the Electropathic and Zander Institute, taking advantage of the high reputation that Zander's work had acquired even in orthodox medical circles.[6] Visitors to the Institute could exercise on Zander's machines, attend a consultation to determine which of Harness' various appliances would be most suitable for their needs, or be given electrical treatment on the spot by one of Harness' assistants. Those unable to visit the establishment in person had only to send a letter describing their symptoms to enable Harness to make a diagnosis of the appropriate kind of electrical appliance required to provide for their needs. Branch offices of the Medical Battery Company were also opened in Leeds and Liverpool. Business was clearly booming and there was plenty of money to spare for glitz and glamour: 'very large sums were spent on advertising the electrical and medical appliances of the company and its methods of treating disease ... the whole of the premises were furnished in an expensive style.'[7]

This was all part of Harness' overarching strategy to build and maintain his own unique brand. The electric belts that they sold were not just any old belts – they were Harness' electropathic belts. Harness was not just any old salesman either, he was the 'eminent electrician' who had invented an electropathic belt with properties beyond that of any other. He also claimed to know things about the way electricity worked on the body that nobody else knew. As one disgruntled and disenchanted former employee of the Electropathic and Zander Institute told the *Electrical Review*:

In my lectures I referred to the labours of Sir William Thomson, Professor Hughes, Erb, and other eminent men. This caused Autolycus [Harness] great annoyance. 'Why should you advertise those fellows,' he would say. 'Don't talk about Sir William Thomson. Tell the public about the great Autolycus, the friend of suffering humanity. Don't talk about general electricity, but say that it is *my* special kind of electricity that cures disease.'[8]

The trick was to make potential patients identify Harness, his Medical Battery Company and their electropathic belts as the only genuine purveyors of electrical relief. William Lynd, the distressed lecturer in this case, provided an insider's view of how Harness ran his business and how he worked hard – and ruthlessly – to build this carefully cultivated image.

Harness played a carefully balanced two-way game as far as the orthodox medical and scientific establishments were concerned. On the one hand his adverts extolled the superiority of his medical and electrical knowledge over that of the rival professionals. They were conservative stick-in-the-muds whilst Harness looked ahead to the future. That did nothing to prevent him from taking full advantage of every opportunity to publish testimonials that appeared to confer the authority of orthodox practitioners on his activities, however. Having acquired the Zander Institute and its equipment, for example, he gave full publicity to an editorial that had appeared in the *Lancet* several years previously applauding Zander's activities, implying that such support extended to his own practices as well.[9] Harness was keen to have his apparatus tested and sanctioned by men known for their expertise and knowledge of medical electricity. He even approached Desmond FitzGerald, former editor of the *Electrician* magazine with the request that he carry out experiments on Harness' electropathic belts. FitzGerald declined, letting Harness know that his 'opinion of your "Electropathic" appliances, and of the methods by which they have been foisted upon an ignorant public, is such as to preclude me from entering into any relations with you such as you propose'.[10] FitzGerald's curt and public refusal certainly did not prevent Harness from trying his luck with other potential experts, and with some success as we shall eventually see. The Electropathic and Zander Institute also organised regular scientific lectures and 'conversazione', imitating elite scientific establishments such as the Royal Institution. Harness was also the founder and President of the Association of Medical Electricians, an organisation established, according to its detractors, for no other reason than to confer an appearance of respectability upon his activities.

As with any other unorthodox or irregular practitioner – and arguably quite a few regular practitioners as well – showmanship was central to Harness' success at attracting customers. Potential buyers had to be convinced of the efficacy of the product by witnessing spectacular displays of electricity in action. One annoyed electrician recorded his attendance at a lecture by an 'eminent consulting medical electrician' (a phrase used often by the press when they referred to Harness, in an ironic play on his own self-aggrandising description of himself) where at the end of the lecture, the power of a Harness electropathic belt was demonstrated by its being used to ring an electric bell and drive a motor. The telegraph engineer J.M.V. Money-Kent described how Harness demonstrated 'belts innumerable and various other wonderful appliances, and also a small motor with a fan mounted on the axis, a bell, and a Holtz machine, all *ostensibly* the invention of this "eminent consulting medical electrician"'. Some of the belts were put to work to demonstrate their superior electricity-generating power. After describing how the belt worked by contact with the skin, Harness duly demonstrated how the one he was wearing could produce a current that worked the equipment on the lecturing table. This was clearly too much for the suspicious telegrapher, who jumped to his feet and accused Harness of having another battery hidden somewhere on his body.[11]

Demonstrations like this were used to assuage any possible suspicions a potential customer might have before they were taken to a consulting room to select their personal electrical appliance. The implication was that if one of Harness' electropathic belts generated enough electricity to put on a good show, they packed enough of an electrical punch to give their wearers their money's worth as well. On this occasion, however, Harness clearly made the mistake of allowing the awkward customer to take over the proceedings. Money-Kent duly found a large battery hidden under the table. In the absence of recalcitrant electricians, this kind of public performance was a highly successful way of getting the Harness message across. It helped, of course, that electrical showmanship like this was perfectly familiar to their potential customers, and was central to the process of bridging the gap between the world of consumerism and the world of science. Watching Harness putting on a show at the Electropathic and Zander Institute was not that different from going to see Professor Pepper perform his spectacular electrical experiments at the Polytechnic Institution. After all, the late Victorian public mainly encountered the products of science in just the same way as they encountered Harness' belts on these sorts of occasions: through exhibition and advertisement.

Harness' basic stock in trade was, of course, his electropathic belts. These comprised a canvas webbing belt into which a number of zinc and copper discs had been sown, connected to each other by wires. The belt was meant to be worn next to the sufferer's skin so that a small electric current would be excited by the wearer's own perspiration. Different arrangements of the zinc and copper discs, connected in series or in parallel circuits, were supposed to produce different therapeutic effects on their wearers' bodies. Harness' promotional pamphlets boasted that the belts were constructed 'on the principle of the galvanic battery known to the great German philosopher Humboldt'. They produced 'a mild, continuous current of electricity, that is specific in some diseases, affords relief in many, improves general health in all, and can never do any harm'.[12] The belts were manufactured in various strengths and sizes, designed to provide the electrical current appropriate for various kinds of diseases. Customers might purchase a 5-guinea, an 8-guinea or even 12-guinea belt according to the severity of the disorder from which they suffered. Jeffery, for example, had clearly purchased one of the cheaper 5-guinea belts. As well as belts, the Medical Battery Company offered appliances designed to fit around wrists or ankles, or to be attached to arms, legs or spines. They all worked on the same general principle.

Harness advertised his belts as being a cure for a whole range of disorders. 'Weak and languid' men were assured that the electric current would bring reinvigoration.

The belts were marketed both at 'nervous, highly susceptible young men who have indulged in malpractices in early youth' and at 'men in the prime of life who have "lived fast"; who have resided long in tropical climates; those who have suffered from malarial fevers or who have been subjected to some great mental strain'. Sufferers like these, particularly the nervous young men, were 'prey to the pestiferous quacks who live on the morbid and unfounded fears their writings, filthy as they are mendacious, excite in the minds of the timorous'. But if they really were impotent, the pamphlets assured their worried readers that 'it may be said, without exaggeration, that the ELECTROPATHIC treatment is absolutely certain, properly directed, to affect a cure'.[13] For women Harness recommended an electric corset, again with zinc and copper discs to be worn next to the skin so that perspiration produced an appropriate current. Pamphlets discreetly hinted that sterile women could be restored to their proper feminine function by wearing one of these. Hernias, rheumatism, hysteria, gout, paralysis, epilepsy and 'female complaints' in general could all be cured by the judicious application of electricity through one of Harness' belts or corsets.

These corsets were a real boon to all suffering women, insisted the Medical Battery Company's advertising literature. The pamphlets promoting them struck a further blow for female education and self-understanding too. Their aim was 'to make the much neglected subject of the disorders incidental to their sex one of popular medical instruction, free from technicalities of language, and strictly decorous in its form … without offending the modesty characteristic of the gentler half of the human race'. The Medical Battery Company's pamphlets argued that lack of knowledge of basic bodily functions along with a misplaced, if understandable, delicacy was making women ill – and even killing them. By making this information available they were acting 'in the interests of purity and national health', since 'the dissemination of sound common sense information upon the physiological attributes of the sexes is one of pressing necessity' that 'cannot fail to be of great benefit to the individual, and will be of advantage to the commonweal'.[14] Women's 'almost total ignorance of a particular function identified with their distinctive sex' was therefore dangerous not only to themselves as vulnerable individuals, but to society as a whole. Harness and his company offered these pamphlets as essential sources of basic sexual information so that 'mothers and guardians of girls' would be 'more capable of properly instructing the young maiden'.[15]

This was a clever strategy on Harness' part. He was trying to pass himself and his business off as committed supporters of some of the most radical and progressive movements in late Victorian society. His electropathic corsets could strike a blow for women's rights. Fulsome encomiums purportedly written by Anna Kingsford filled several pages of Harness' pamphlets, congratulating him on his 'wise advice' to women and recommending his belts for cases of '*painful* hysteria'.[16] Kingsford was a committed campaigner for vegetarianism, the anti-vivisection movement, women's rights and spiritualism. She advocated physical and mental exercise for women in pursuit of a healthy mind in a healthy body, just as the proponents of muscular Christianity for men did. Unable to get the training she wanted in England, Kingsford had studied medicine in Paris and qualified as a doctor in 1880. Her spiritualism and anti-vivisectionism sat uneasily with the perspectives of late nineteenth-century Parisian medicine. She even claimed to have used her mental 'spiritual thunderbolt' to cause the deaths of the French physiologists Claude Bernard and Louis Pasteur – committed advocates of animal experimentation.[17] Kingsford was already dead by the time her words and reputation were appropriated by Harness to help sell his belts and corsets. His choice of her as champion says a great deal about the kind of women that the Medical Battery Company expected to read their literature and buy their products.

The scale of Harness' activities and the efforts devoted to maintaining his position are a clear indication of the popularity of medical electricity as he practised it. Without a market ready and willing to accept his claims of expertise and the effectiveness of his electrical appliances in curing their ills, there could have been no Harness and his commercial empire. His opponents explained the success of his endeavours by invoking a combination of public gullibility and Harness' own unscrupulousness. He was portrayed as preying upon the weakest and most vulnerable members of society. Ascertaining just who exactly Harness' customers might have been is not an easy task – after all we do not even know who Mr Jeffery was beyond his occupation as a bank clerk. His detractors almost without exception sought to portray those who took advantage of his services as gullible, ignorant and often poverty stricken. It is clear, however, particularly given the prices Harness charged for his services and products, that most of his patients were broadly from the middle classes. Even the 5 guineas that Jeffery paid, or rather didn't pay, for his electropathic belt was a significant sum of money. What his customers thought of Harness' products is even more difficult to recover. During the course of their campaign against him, the *Pall Mall Gazette* published dozens of letters from disgruntled customers, complaining about the way in which Harness had duped them and congratulating the *Gazette* on its campaign of vilification. Nothing less would be expected, of course, but the letters at least suggest that a significant portion of those who bought Harness' products, far from being silent sufferers, were eloquently literate, informed and vociferous.

Harness' electropathic belts, corsets and other electrical medical impedimenta clearly managed to reach a clientele that orthodox medical practitioners could not satisfy. The glamour of electricity presumably provides one explanation for the popularity of his appliances. His customers knew that they were living in an age when new electrical technologies of all kinds were proliferating, and for which the sensuous tactility of electricity was already an established fact.[18] There was a long history of electrical showmanship surrounding the body into which Harness and his products could fit quite neatly and on which he certainly drew when making his pitch for business.[19] People living in an age that placed value on mass-produced commodities were easily attracted by the prospect of managing their bodies through the products of mass-production as well. Orthodox medical practitioners themselves were not above suspicion of gulling the public in any case, quite often being portrayed as attempting to maintain a monopoly on healing. This was certainly the view promulgated by Harness and his like. Electric belts were quite certainly not the only alternative therapy flourishing during the closing decades of the century either. Jeffery might have visited a homeopathic practitioner or a hydropathic establishment instead. Middle-class Victorians' bodies were becoming battlegrounds for different groups trying to establish their credentials as managers of individual and national health.

The courtroom battle in which Harness and Mr Jeffery found themselves embroiled in 1892 – and the series of court cases that followed over the next few years – give us an insight into how electricity fitted into wider anxieties about individuals, nature, society and science at the end of the Victorian era. The electrical and medical professional press had been spluttering about Harness' activities for a number of years. The vulnerability of his position, now exposed by Jeffery's stubborn refusal to cough up the cash, encouraged the populist *Pall Mall Gazette* to weigh into the argument as well. From October 1893 the newspaper ran a series of exposés on Harness and his activities, revealing him as a cad and a charlatan. By the end of the decade, Harness was broke and the Medical Battery Company in receivership. Jeffery matters, therefore, despite his obscurity, because he provided the momentum that set this particular electrical *cause celebre* in motion. The trials dragged in all sorts of people, they even featured the aging William Thomson, Baron Kelvin, doyen of Victorian physics, as an expert witness. They show, therefore, how by the end of the nineteenth century bodily electricity was thoroughly embedded in popular culture. The next few chapters will survey electrical entrepreneurship about the human body during the final quarter of the century, following debates about how electrical bodies were meant to work, who held rights over the electrical body and what that tells us about science at the end of the Victorian era.

FOURTEEN

ELECTRIC ENTREPRENEURS

When Mr Jeffery walked into the Medical Battery Company's plush apartments on Oxford Street in 1892, he was making a conscious choice. Obviously, he was deciding that he wanted to try an electrical cure for the physical ailment that was worrying him. But beyond that, he was also making a decision about what kind of electrical treatment he wanted and what kind of medical electrical practitioner he was willing to trust. Medical electricity was contested territory. Purveyors of electrical paraphernalia such as Cornelius Bennett Harness were by no means the only ones trying to occupy that potentially fertile and lucrative ground. In 1887 the *Electrician* announced that electricity 'as a therapeutic agent in medicine and surgery has now gained an acknowledged place'. No longer was it 'apt to be looked upon … as the practice of an occult art'. Electrical medicine was now an 'exact science'. The *Electrician*'s editor did not care whether 'the actual way in which the electric current works on the human system' was known or not. What mattered was the 'exact method of computation', which meant that 'the operator, armed with one of the new instruments with all its modern improvements, may regulate at his will, graduate the flow, and operate with the utmost precision'.[1]

Some years later, the *Lancet* agreed with its electrical counterpart on the question of medical electricity's new-found reputation. 'The services which electricity can perform in medicine and surgery are both numerous and varied', decreed their article.[2] By the middle of the 1890s the medical marketplace in electrical cures was flourishing. Electrical departments were increasingly common in large metropolitan general hospitals; they could be found in specialist hospitals devoted to treating specific diseases and at least some general practitioners used electrotherapy in their practices. Even electrical engineers were taking an interest in medical electricity. If, as the *Lancet* had it, electrotherapeutics in the past had been 'singularly unfortunate in that it has been grossly abused by people who have made of it a stalking horse for rampant quackery', there was now a 'fresh interest in medical electricity' driven by 'the impetus given to the study of electricity by its commercial applications'. It was time to take electrotherapeutics out of the hands of 'the self-styled unqualified "medical electrician", male or female' and put professionals in charge. As the *Lancet*'s editorial indicates, competing groups of electrical entrepreneurs were battling over the electrical body throughout the final quarter of the nineteenth century. Hospital doctors, irregulars like Harness and enterprising electrical engineers were all vying with each other for the privilege of Mr Jeffery's custom.

In 1892, whilst Jeffery was struggling with his electropathic belt, Henry Lewis Jones at St Bartholomew's Hospital could boast that 'our Electrical Department is one of the most important in this country'.[3] The electrical department had been established a decade earlier under the aegis of William Edward Steavenson. Steavenson was himself a St Barts' old boy, having studied at the hospital's medical school before heading to Cambridge to acquire his MD in 1884. It was a significant responsibility for a rising medical star, and a chance to make a reputation in what must have seemed to be a promising and highly visible new field of scientific medicine. The new department was closely modelled on the pioneering electrifying room at Guy's Hospital, where Golding Bird had worked almost half a century earlier. No doubt the department was rather less luxurious in its fittings than its would-be competitor on Oxford Street. It was built on the site of the old coroner's court at the hospital, conveniently isolated from the main wards so that everyday comings and goings would not interfere with the sensitive apparatus installed there. There were two main treatment rooms with electrical equipment, wired up to a series of batteries kept down in the hospital's basement, and there was a separate room for patients to be treated by an electric bath.[4]

Hospitals other than St Bartholomew's were also keen to develop and expand their own facilities for electrical medicine. Following Steavenson's own tragically early death in 1891, Henry Lewis Jones took over his position there and continued the work of the electrical department, intent like his predecessor on making a name for himself as a scientific doctor. Charles Bland Radcliffe, physician to the Westminster Hospital from the late 1850s, and William Henry Stone, who was appointed physician at St Thomas' Hospital in 1870, were both well-known practitioners of electrical medicine as well as being prolific publishers on electrical matters. Armand de Watteville, author of the *Practical Introduction to Medical Electricity*, was in charge of the electrical department of St Mary's Hospital as well as being electrotherapeutical assistant to University College Hospital. De Watteville, like many of this new generation of pioneering electrical doctors in the large metropolitan hospitals, keenly advocated applying the latest developments in electrical technology and practice to electro-medicine. In particular they emphasised the importance of measurement and accurate, standardised doses of electricity. This was what the *Electrician* meant when it talked about the importance of 'exact science'. These men straddled the worlds of electricity and medicine; many of them had links with the Society of Telegraph Engineers as well as the more usual and medically orthodox connections with the Royal College of Physicians or of Surgeons. As we shall see later in this chapter, this was a fine balancing act, and even men with these kinds of impeccable backgrounds in respectable, professional and orthodox medicine could not always manage it without falling off the tightrope.

How did these hospital electrical departments operate, and who were their patients? When St Barts' appointed Steavenson as their medical superintendent in charge of the new department, they also hired an electrician who would be responsible for the day-to-day business of running the department and keeping equipment in order:

> [t]he man employed to make these repairs will also have the custody of the building in the absence of the medical officer. He will receive notices from the wards of patients requiring electrical treatment by the electrician, and supply and receive back again the batteries used in the wards by other medical officers.[5]

It was a simple enough organisational model: doctors prescribed and oversaw the treatment; electricians got their hands dirty. In-patients were usually referred to the electrical department by the doctor in charge of their treatment on another ward. If St Barts' was typical, then there were an even larger number of out-patients who came to hospitals specifically for electrical treatment. The case notes at St Barts' contain too little information to tell us anything very useful about the patients' backgrounds. Just under 40 per cent of the in-patients seen at St Barts' in 1886 were women. A fair number of the out-patients were receiving treatment for what might loosely be described as occupational illnesses. There were twenty-nine cases of lead palsy, for example. By far the largest number of cases seen as out-patients, however, were the 100 instances of infantile paralysis.[6]

Electrical practitioners in the large metropolitan hospitals quite clearly saw themselves as part of the future. They had cast off the shackles of quackery and plainly regarded themselves as the vanguard of scientific medicine. A new age was dawning for electrotherapeutics. William Snowden Hedley, former army surgeon and the medical office in charge of the electrical department at the London Hospital, could afford to shudder appreciatively as he recalled:

> the coarse instrumentation and crude procedure of a comparatively recent past, when an unmeasured current was 'dabbed' on to the body at the end of a couple of sponges, and it was orthodox practice to administer faradic torture by means of a couple of metal cylinders grasped by the patient.[7]

Everyone now knew that 'an increased attention to electrical physics' was 'the only sure foundation for electrotherapeutics'. There were other signs that electricity's time had come – 'more formal recognition by the chief metropolitan hospitals' amongst them. By 1900 Hedley was one of the founding editors of the *Journal of Physical Therapeutics*. A few years later, the British Electrotherapeutic Society was established and the British Medical Association formed a sub-section for medical electricity. These were the trappings of professional prestige and respectability. They were the signs of institutional success.

Medical electricity was flourishing in specialist hospitals too, and practition-ers there were just as keen to advertise their professional credentials. Medicine at the end of the nineteenth century was a highly competitive affair, and anything that offered an edge was worth pursuing. Back in 1866, doctors at the National Hospital for the Paralysed and Epileptic boasted about their electrical room and the *Lancet* hailed it as 'a model of its kind'. It was 'the only room of the sort in English hospitals which justly represents the present condition of electrical sci-ence as applied to medicine'. It boasted a 'Muirhead's battery of 100 cells, arranged especially for medical use by Messrs. Elliott, Brothers' and offered 'periods espe-cially devoted to electro-therapeutics' for all comers on 'the afternoons of Monday, Tuesday, Wednesday, and Friday'.[8] Ambitious doctors stacked up physicianships at specialist hospitals both as valuable assets in their own right and as stepping stones to the even more prestigious general hospital positions. Herbert Tibbits listed positions at several such institutions on the title pages of his books. Like Tibbits, William Steavenson listed an affiliation with the Hospital for Sick Children on Great Ormond Street amongst his own portfolio of covetable placements. One important reason for seeking appointments at these sorts of institutions was the opportunity they offered of acquiring valuable patronage. Specialist hospitals typically had long lists of aristocrats, politicians and wealthy socialites on their governing boards.

General practitioners were turning to electricity as a potentially useful weapon in their armoury too. One advocate declared that a 'medical man engaged in general practice can utilize electricity in a number of different ways'. He argued that almost 'every known form in which electrical force can be exhibited finds its application in some special department of medical or surgical work'.[9] From the 1840s, if not earlier, electrical instrument-makers had regarded doctors as potential customers for electrical instruments, designing batteries, magneto-electric devices and induction coils for a specifically medical market. By the 1890s medical electrical instruments for the general practitioner were quite easy to come by. They were usually sold in a small, compact wooden box which contained a drawer for keeping the elec-trodes, some spare wires and a small beaker for measuring the acid for the battery. Apparatus like this could, quite literally, fit into the pocket of a travelling, working doctor's greatcoat. The invention of dry cells was a major boon in this respect:

> These coils have got the advantage that there is no liquid required to work them. As it is only the spilling of acid which makes a coil go out of order, these coils require practically no repairs and less attention, and are cleaner, more reliable and conven-ient than the old acid cells.[10]

These kinds of small, portable apparatus typically sold for between £2 and £3 each and were well within the financial reach of all but the most impecunious family doctor.

Herbert Tibbits provides us with a good example of how affiliations with specialist hospitals could help build a medical career, and provide ambitious doctors
with opportunities both for marketing electrotherapeutics and for advertising
their expertise. When the West-end Hospital for Diseases of the Nervous System,
Paralysis and Epilepsy was founded in Welbeck Street in 1878, the prime mover
was Herbert Tibbits. As we have seen, Tibbits, a fellow of the Royal College of
Physicians of Edinburgh as well as a member of the Royal College of Surgeons, had
previously been medical superintendent at the National Hospital for the Paralysed
and Epileptic in Bloomsbury, as well as medical officer for electrical treatment at the
Hospital for Sick Children, Great Ormond Street. Tibbits had already made quite
a name for himself as an authority on electrical medicine and had been publishing
prolifically on matters electrotherapeutic for a decade or more. Before branching
out with the establishment of his own specialist hospital he had been responsible
for overseeing the administration of electrotherapy at the National Hospital, an
institution modestly described in his own words as 'the pioneer and outpost, so to
say, in this metropolis of the scientific and methodical application of electricity to
the alleviation and removal of disease'.[11]

In 1886 Tibbits attempted to add to his fledgling electrotherapeutic empire by
establishing the West-end School of Massage and Electricity in conjunction with
the hospital. By then he had acquired something of a reputation as a particular
advocate of the use of electricity in connection with massage as well. The particular aim of the school was to provide training in electrotherapy and in massage for
nurses. Tibbits boasted that:

> our Hospital stands quite alone … in having associated with it a School attended
> by numerous Students … who are carefully and thoroughly trained, who attend
> daily to Massage, Electrize, and apply other methods of local treatment as directed
> by, and under the supervision of the Medical and Surgical Staff, and who, at
> the end of their training, and after passing a written and practical examination,
> receive Certificates that they are 'competent to apply Massage and Electricity under
> Medical Authority'.[12]

Tibbits was assisted at the school by Thomas Stretch Dowse, another advocate
of the joint use of electricity and massage. The two of them had developed a
system of massage in which the patient would have electricity directed towards
particular parts of the body whilst they were being massaged, with the masseurs
or masseuses themselves being the means of conducting the electric current into
the patient's body.[13]

As we shall see quite clearly in the final chapter of our exploration of Jeffery's
world, Tibbits actually inhabited a rather murky border country between reputable
and disreputable medicine. However much the champions of hospital-based electrical medicine such as Hedley liked to trumpet their complete divorce from dubious

practices that might bring electrotherapy into disrepute, the truth was rather more slippery. There was, to say the least, some shadowy ground in between where apparently orthodox and respectable medical electricity shaded into the heterodox. Tibbits was certainly not the only late Victorian doctor whose quest for fame, as well as the simple need to make a living from medicine in straitened circumstances, led them to this dangerous borderland. Harry Lobb a few decades earlier had found himself in similar circumstances. Just what counted as orthodox and respectable practice was often open to debate. Even so innocuous a matter as publishing a book could in some circumstances be regarded as unprofessional behaviour by the governing councils of the Royal Colleges of Physicians and of Surgeons. Advertising in any form at all was considered utterly beyond the pale. This meant that enterprising, entrepreneurial and ambitious electrical doctors had to tread very carefully indeed as they tried to make names for themselves. A foot out of place could easily see them consigned to the outer darkness of their profession.

By the 1880s there was certainly an ever increasing number of companies registered as dealing in electrical medical appliances of various kinds. On paper, at least, they looked like highly lucrative propositions. In 1887, for example, the Electric Medical Appliance Company Ltd was established with an announced capital of £2,000 in £1 shares, with the object 'to deal and trade in electrical appliances, trusses and fancy articles generally in the United Kingdom and the United States of America'.[14] At about the same time the Medical-Electric Belt, Truss, and Health Appliance Company Ltd was established. They claimed a capital of £20,000 in £5 shares. Their object was to take over the business of E. Fava & Co., who were already established as 'manufacturers of electric belts and batteries, trusses and health appliances &c.'.[15] The Pall Mall Electric Association was another frequent advertiser of medical electrical products. Interestingly, magazines such as the *Telegraphic Journal & Electrical Review*, aimed at the burgeoning electrical professions and potential investors, included concerns such as this in their lists of new electrical companies registered without comment. By far the most common appliances seem to have been electric belts. More unusually a belt might include means of attachment to an additional, external battery. In addition to belts and corsets, electric combs and even hairbrushes were offered for sale, along with electric towels and a whole battery of electric pills of all kinds. The companies also offered free consultations so that potential patients might ascertain which of their range of products might be most appropriate as a cure for their ailments. For those unable or unwilling to be seen in person, consultations were also usually available through the post.

Irregular practitioners were quick to point the accusing finger of quackery at each other. T.G. Hodgkinson complained that medical electricity was beset by quacks:

In the present age, as in the past, the charlatan has done much to cover a noble science with ill-repute - squirming the ink-like fluid of ignorance and effrontery -

discoloring [sic] the clearer water of intellectual inquiry. Small wonder that under these circumstances, thinking men and women of culture have regarded the practice of electropathy as an 'unclean' thing, and rather than risk defilement by actual association, have cried 'Away with it'.[16]

The preferred antidote to quackery in this case, however, was Hodgkinson's own electro-neurotone apparatus, which he succeeded in having puffed in no less prestigious a place than the pages of the *Lancet*. William Johnson, a former employee both of J.L. Pulvermacher and the Pall Mall Electric Association, presented himself as having:

a distinct advantage over those of my contemporaries, who, having embarked on what they are pleased to term a scientific career from a purely commercial standpoint, have pursued their devious ways, actuated solely by the sordid money-making instinct, and utterly regardless of, if indeed they are not altogether unacquainted with, the nobler aspirations which should animate the true scientist.[17]

E.S. D'Odiardi, extolling the virtues of his own electrical equipment, also suggested that he had been responsible for 'finding out the true principles, which are likely to become the pole-star of medical electricians'.[18]

The list of diseases that irregular practitioners argued in their puffs and advertisements could be cured by electricity was impressive. The various devices they sold could:

give wonderful support and vitality to the internal organs of the body, improve the figure, prevent chills, impart new life and vigour to the debilitated constitution, stimulate the organic action, promote circulation, assist digestion, and promptly renew the vital energy the loss of which is the first symptom of decay.

Advertisements promised that electric belts alone could 'stimulate the functions of various organs, increase their secretions, give tone to muscle and nerves, relax morbid contractions, improve nutrition, and Renew Exhausted Nerve Force'.[19] Electric towels promised a 'cure for rheumatism, neuralgia, constipation, indigestion, liver complaints, &c.'.[20] Whilst the electric hairbrush 'Prevents falling hair and baldness! Cures dandruff and diseases of the scalp! Promptly arrests premature greyness! Makes the hair grow long and glossy! Soothes the weary brain!'[21] Many of these advertisements for medical electrical products generally hinted, circumspectly or otherwise, that the electricity produced by the various appliances could also restore the sexual debility brought about by masturbation or excess, and they showed manly men and womanly women restored to perfect function through the powers of electricity.

From at least the mid-nineteenth century onwards an increasing number of medical authors had bombarded the public with concerns over the physical dangers of sexual incontinence.[22] Popular writers such as William Acton repeatedly emphasised the degree to which uncontrolled intercourse or masturbation could cause permanent damage to an individual's health in later life. Men were putting themselves at risk of impotence or spermatorrhaea. Women risked nymphomania, hysteria or even sterility if they engaged in uncontrolled sexual activity.[23] The medical model underlying, or at least bolstering, these claims was one in which the body was held to contain only a limited amount of nervous energy. Depletion of this limited supply through excessive sexual activity was the cause of nervous disorders of various kinds. The newly promulgated doctrine of the conservation of energy in the physical sciences provided strong support for such views of the body as we have seen.[24] This is one reason why electricity was regarded as a cure for such diseases. Promoters of electrical health appliances designed to restore virility argued that the electricity their products produced, replaced the missing nervous energy, thus restoring the patient to vitality. Of course, electricity had a long history of being marketed in this fashion, stretching back at least to Graham and his celestial bed.

James Beresford Ryley was one doctor on the fringe of orthodoxy who made great play with the claim that electricity was the cure for self-induced sexual dysfunction. 'Sexual incompetency and irritation' was 'one of the commonest symptoms associated with all forms of physical and nervous exhaustion'. It was 'often due to self-abuse in early life, or excessive venery later on'.[25] Electricity could restore the flagging tissues: 'By the one quality it removes the products of inflammation … and by the other it operates on those genetic forces the suspension or diminution of which constitute complete or relative impotency.'[26] Such a course of treatment could revolutionise the patient's entire self-image. Describing the response of one patient to electric treatment, Ryley related how:

> He had such pride in his physical development while under treatment, that for some time after he invariably invited his friends to feel his chest and biceps muscles, and to that end, struck an attitude, à la Sandow, and at the same time doubled up his arm in the usually approved manner for that purpose.[27]

Electricity could produce the body beautiful by overcoming the deleterious effects of culture and lack of self-control. Women could benefit as well. Ryley recommended 'electro-massage' for 'chronic womb-diseases' and as a cure for sterility.[28]

The increasing physical degeneration of the human race was widely regarded as a growing problem by the end of the nineteenth century.[29] Electricity was only one of a number of technologies that could be represented as potential rectifiers of the condition. Gymnastic machines, diet and new fitness regimes were all advocated as ways of returning the civilised human body to its past condition. Ryley's view was that:

Irrespective of the Biblical account of the creation of man, there is every reason, on scientific data, to believe that he has greatly degenerated physically and mentally with the growth of civilization, and that new diseases unknown and unfelt in earlier times have been added to the list of his sufferings.[30]

Hodgkinson concurred with this estimation, asserting that the 'present generation presents undoubted evidence of degeneration in nerve power' and that a 'consensus of opinion has selected electricity as the best weapon' for combating the trend.[31] The pace of modern life, the mental stresses of *fin de siecle* culture and the sedentary occupations of the late Victorian middle classes were all variously cited as causes of this precipitous physical decline. Unfit bodies led to unfit minds, and electricity offered a sure way of stopping the rot.

William G. Johnson offered a wide range of custom-made belts and other medico-electrical devices from his premises at 64 Cambridge Terrace. Though as we have seen, he still advertised himself as 'fourteen years manager to the late J.L. Pulvermacher, and Manager and Consulting Electrician to the late Pall Mall Electric Association, and the Medical Battery Company', he professed disdain for 'self-styled "eminent consulting Electricians"' (a clear reference to Cornelius B. Harness). He boasted that he had 'recently succeeded in elaborating and perfecting an Electric belt, at once Hygienic and Electro-therapeutic'.[32] His belts for ladies and gentlemen alike could cost as much as 63s – and as little as 10s 6d. Sufferers could buy electrical lung protectors for 15s, electrical knee caps for 5½s and magnetic soles for the boots at 7s 6d a pair. T.G. Hodgkinson also offered his electro-neurotone apparatus as an antidote to 'the miserable record of shameless imposition upon the part of those, who by means of so-called Galvanic Belts, and similar appliances, rob and deceive suffering humanity'.[33] Hodgkinson's invention worked on the principle that the body was like 'the "Exchange" of a large telephone system'. The brain was the switchboard and the 'Great Sympathetic Nerve' was like 'the telephone cable into which many conducting wires are placed for further distribution to special series of sensory and motor nerves'.[34]

Joseph Granville Mortimer's invention of the electro-mechanical vibrator was similarly based on a very specific view of the way the nervous system operated. The nerves operated by transmitting vibrations through the body and imbalances could be brought about by causes ranging from mechanical shock – as in the case of 'railway spine' – to sexual excess. His vibrator, or percuteur as he first called it, worked by propagating its own vibrations through the system. Mortimer argued that all 'nerve action, both normal and abnormal, consists of vibrations, and that *other* mechanical vibrations set up and propogated by any mechanical apparatus or appliance' could 'modify the internal movements' and restore normal nerve function.[35] Mortimer's earlier versions of the instrument worked by mechanical clockwork, but later versions used electricity to produce the necessary vibrations, though he was keen to emphasise that electricity played no direct role in their operation, it was simply a means of producing the appropriate vibrations. The vibrator could either be applied directly on the affected area of the body 'with varying force and speed until alleviation of the trouble is produced' or 'the percussion may be made higher up in the trunk of the nerve, over the centre, or at a more distant point in the periphery'.[36]

Just as Mortimer was pushing his new instrument, a new form of treatment for hysteria was being imported from the United States. In uterine massage, the:

> cervix uteri is first pulled gently forwards and allowed to fall back six or more times; the uterus is then gently squeezed between the two hands in all its parts … the uterus should also be frequently pushed up and held up in the pelvis by the finger in the vagina.

The whole process was to take between ten and thirty minutes depending on the patient's response to the stimulation. The *Lancet* took a dim view of this new American import, passing disparaging remarks regarding the 'personal so-called magnetism' of its practitioners.[37] It was not long before both practitioners and their female patients realised how useful Mortimer's vibrators might be in this respect. Mortimer was unamused. 'It will be monstrous if this method is reduced to the level of a "cure",' he spluttered, 'by placing percuteurs in the hands of patients and allowing them to use them for themselves.'[38] To make his point clear, Mortimer insisted that he had:

> never yet percussed a female patient … I have avoided, and shall continue to avoid, the treatment of women by percussion, simply because I do not want to be hood-winked, and help to mislead others, by the vagaries of the hysterical state or the characteristic phenomena of mimetic disease.[39]

He was not the first doctor, and certainly not the last, to find patients using their instruments in ways that they had not quite anticipated.[40]

Not only electrically minded medical practitioners were looking for new ways and new instruments to apply electricity to medicine, professional electricians and electrical engineers were themselves also taking an increasing interest in the profitable application of electricity to the human body. In his inaugural address as the new President of the Society of Telegraph Engineers and Electricians, Lieutenant Colonel Charles Webber called on doctors to join the society. 'The members of the medical profession should be induced to join the Society,' he argued, so 'that their knowledge in the curative applications and principles of electricity might be perhaps increased, for the day was approaching when the study of electricity in its physiological function would be of paramount importance to medical men'.[41] William Henry Stone, physician and electrician at St Thomas' Hospital and a fellow of both the Royal College of Surgeons and of Physicians (the first, in fact, to hold both fellowships) responded to Webber's call. Himself a member of the Society, he told his audience there how they could:

> be of very great service to us physicians and physiologists by giving us suggestions, especially in the department with which you are most conversant (more conversant than we are), viz., the department of measurement. Medicine and its kindred departments lend themselves extremely ill to measurement. The tone of mind required in the physician is not that of the mathematician – it is rather judicial than computative. It is often a question of weighing doubtful evidence and of balancing alternatives rather than of solving an equation [42]

It was a call to bring electrotherapeutics under the purview of electrical engineers.

Before the end of the 1880s Webber and Stone's challenge was accepted with the foundation of the Institute of Medical Electricity. The Institute was founded in 1888 as a joint-stock company by a consortium of electricians, including many prominent members of the Society of Telegraph Engineers and Electricians (now renamed the Institute of Electrical Engineers).[43] The shareholders included, amongst others, Silvanus P. Thompson, whilst the chairman of the Board of Directors was William Lant Carpenter, son of the eminent physiologist William Benjamin Carpenter. As well as Thompson, the Institute's visitors included Sir John Lubbock, Professor William Ayrton and William Henry Preece, the Post Office's Director of Telegraphs.[44] The Institute's managing director and resident electrician was Henry Newman Lawrence, himself a member of the Institute of Electrical Engineers who had been trained in electricity by William Ayrton at Finsbury Technical College.[45] The resident physician was Arthur Harries. The Institute's aims were to 'provide establishments where Electrical Treatment may be obtained under the direction of qualified medical men at moderate fees', to 'encourage by scientific investigation and other approved means the use of Electricity in its applications to Medicine', and to 'provide medical men with the use of reliable electrical apparatus and instruments for diagnosis or treatment of Patients'.[46] Another, unstated, aim was to make money for the shareholders. Members of the Institute of Electrical Engineers clearly regarded applying their expertise and technologies to the electrical cure of the body as a potentially highly lucrative enterprise.

The Institute of Medical Electricity started life in a building on the corner of Regent Street and Coventry Street, but soon moved to occupy premises on Fitzroy Square, a stone's throw away from the prestigious medical practices of Harley Street. Its proprietors promised to deploy a panoply of technologies in the service of its patients. Visitors and patients would find themselves in 'a large and cheerful-looking room … in which, upon the wall, are fixed switch-boards arranged with suitable connections for supplying all the forms of current used in ordinary cases'. There were separate rooms for treating male and female patients.

In the bathroom was 'a large, well insulated oak bath, having connections conveniently arranged around it … Sponge, plate and bar electrodes of various sizes
are used for the different form of electric bath here administered, and all are
ready to hand'. Following their visit, the editors of the *Electrician* were certainly
impressed: 'High-class instruments, skilled attendants, accurate measurements
and careful records. We can hardly imagine an establishment better fitted for the
difficult and unique work the directors have set themselves, nor managed in a
manner more worthy the support of medical and scientific men, not to mention the public generally.'[47] Newman Lawrence's careful wooing of the press had
clearly paid dividends.

The Institute's centrepiece was the 'Electrified Room System', an 'insulated
metal floor and an insulated metal ceiling, each of which is connected to one
pole of the secondary circuit of a powerful induction coil'.[48] Patented by Captain
Arthur Byng, the arrangement was designed to set up a powerful and rapidly alternating electric field between the floor and the ceiling so that if any conductor
(such as a human body) were placed between them, a rapidly alternating current would be induced in that conductor as well as a static charge. To complete
the treatment, the patient placed within the field would have a massage administered by an earthed attendant so that a current would flow to earth from the
patient, through the hands of the masseur or masseuse. The Institute also featured
electro-gymnastic apparatus, upon use of which a patient could have electricity
administered whilst carrying out a range of exercises. Through a combination of
electrical and gymnastic treatment patients were assured that 'weak or injured,
and even paralysed limbs may, as long as their vitality is not entirely destroyed, be
rapidly restored, and that the process is simple, natural and pleasant'.[49] A regime
of treatment like this would cost the patient 5s for a half-hour session, as would
electric manipulation in the electrified room.

Whilst the Institute of Medical Electricity's proprietors clearly did not regard
themselves as quacks, others were not so sure. The Institute was soon in difficulties
with the Royal College of Physicians over its efforts to employ some of the college's members as physicians. In April 1888 the college's Board of Censors received
a letter from William Steavenson, asking their permission to take up the post of
physician to the Institute. He assured them that 'it appears to be a bonâ fide and
respectable company, and if properly managed will supply a great want, if it is sup-
ported by men of recognized position and good repute'. He wanted to know 'if
the Royal College of Physicians would view with disfavour the acceptance by a
Member of the College of such a post as the one now offered to me' and he made
it clear that he had:

> told the representatives of the Company that I could not allow my name to be
> advertized in the lay press; that I should require full ability to accept for treat-
> ment or reject any patient not sent to the Institute by a medical man; and, that I

could not accept the post if such a step were disapproved of by the Royal College of Physicians.[50]

His letter was to no avail; the Board of Censors did indeed disapprove, just as they disapproved of William Henry Stone taking up a similar appointment offered by the Institute.

The college's Board of Censors, despite numerous pleas from the Institute of Medical Electricity, remained adamant that its members could not accept employment at such an establishment. Their view was that it was 'unprofessional for a Member of the College to ally himself with a commercial company formed for the treatment of disease'. Letters begging the college to change its mind, signed by leading electricians and doctors respectively, had no effect. Newman Lawrence even wrote to them asking that the Council 'appoint a representative of that body to visit this Institute, to enquire into its methods of working, and to make a report thereon to the authorities of your College at his discretion'.[51] The offer was brusquely declined. This is a good example of the shadiness of the boundary between quackery and orthodoxy that bedevilled electrical entrepreneurs, even those who tried their damnedest to stay on the right side of respectability. Whilst Newman Lawrence and his co-workers at the Institute of Medical Electricity saw themselves as white knights, combating quackery by establishing a foundation devoted to the strictly scientific application of electricity to the body, the Royal College of Physicians regarded their participation in trade and willingness to advertise their services directly to the public as symptomatic of it. To the untutored or jaundiced eye there clearly seemed little difference between the therapies on offer at the Institute and those that might be purchased from any other vendor of medical electrical appliances.

So there was certainly plenty of choice facing Mr Jeffery as he pondered where to go for electrical treatment. Furthermore, there would have been relatively little difficulty in finding out about such treatments. New magazines and newspapers proliferated, reaching an ever expanding audience, and the new mass market and the development of new printing technologies during the nineteenth century wrought massive changes in the technology of advertising by the final quarter of the century.[52] Whereas during the 1830s or 1840s an advertisement for a medical electrical product would have consisted of little more than half a column of dense print on the final page of a magazine or newspaper, by the end of the century large, fulsomely illustrated advertisements were commonplace. Notices for electrical medical appliances took their place in this expanding print media along with a whole array of others. Advertisements for pills, lotions, tonics, exercise machines and health foods jostled with each other in the pages of newspapers. Nevertheless, there is little doubt whose presence was most visible. Jeffery might have had to delve deeper to find out about the Institute of Medical Electricity, or to discover Steavenson's or Stone's medical electrical credentials. Harness, on the other hand, was everywhere. It would have been quite clear to anyone who trusted what they read in the papers that the 'eminent consulting Electrician' was the man to trust when it came to applying electricity to the body. If publicity was what mattered, then there was no contest. Equally, there was plenty of other negative material in the press that might have made an intelligent man think twice before deciding to submit himself to the current.

FIFTEEN

MEASURE FOR MEASURE

Did Mr Jeffery play cricket? We do not know for sure, of course, but it is at least reasonably plausible that he did. He belonged to just the right aspirational social class that was meant to enjoy virtuous and health-giving Sunday recreation. Who knows – maybe it was on the cricket field that he acquired the unfortunate hernia that eventually led him into Cornelius Bennett Harness' clutches. If he paid any attention at all to cricketing gossip he might well have come across and wondered at the story of the cricket-playing soldiers and their close encounter with electricity on the playing field. That particular incident took place on a warm summer's day in 1890 in Aldershot, as groups of soldiers lounged around beneath some trees at the edge of the field, watching some of their comrades playing a game of cricket. We know that it must have been a heavy, humid afternoon since later on in the day, the storm clouds gathered. There was 'a sudden flash of lightning, like the bursting of a shell … followed by cries for help from the men under one particular tree'. Five men had been simultaneously struck by lightning. One of them was dead. The second had stopped breathing, but still registered a faint pulse and eventually came to after artificial respiration, and the remaining three were out cold, but recovered consciousness without any other help within a matter of minutes. The four survivors were back on duty within a few days.[1]

This reminder that electricity could deal death as well as life might well have caused a mild flutter of unease in anyone contemplating experimenting with an electrical cure. There was, of course, another reason why death by electricity was in the news during the summer of 1890. On 6 August in Auburn Prison, New York, the unfortunate William Kemmler became the first victim of electrical judicial murder. His botched execution caused outrage and fractious debate on both sides of the Atlantic. Was this a civilised and scientific form of execution as its supporters argued, or was it a barbaric and unseemly perversion of electricity's powers, as the many detractors of electrocution (a new word for a new kind of killing) averred? Kemmler's end certainly raised quite a number of uncomfortable questions for electricity's advocates. Kemmler's fatal electrical encounter put in stark relief the matter of electricity's relationship with life, just as Weems' post-mortem galvanisation had a little over seventy years previously. It cast a rather uncomfortable light as well on what was, for the Victorians, a rather murky hinterland between the lands of the quick and the dead. Electricians were starting to find themselves in the strange and rather unexpected position of being asked to police the borders between life and death. During the 1890s they started to develop tools that would allow them to do just that.

The Holy Grail of late Victorian physics was accuracy. 'Accurate measurement is an English science' boasted the Irishman and adopted Scotsman, William Thomson. Metrology – the science of measurement – was itself a nineteenth-century invention. Thomson – Baron Kelvin of Largs by the 1890s as a result of his efforts in the telegraph industry – thought that measurement was an essential part of the process of making electricity real and useful. Measuring and standardising units of electricity would make it a 'real, purchaseable, tangible object' and Thomson looked forward to the day when one could 'buy a microfarad or a megafarad of electricity' just like a pint of milk. The great iconic experiment of late Victorian physics was an effort to do just that by defining a standard unit of electrical resistance – vital for Britain's all-important telegraph industry.[2] Measurement was about defining the new discipline of physics as well. Think of the clockwork punctiliousness of Phileas Fogg, gently parodied in Jules Verne's *Around the World in Eighty Days*; self-discipline was part of what it meant to be a Victorian gentleman. The art of measurement could be thought of as a way of teaching that gentlemanly self-control. Many physicists and electricians were keen to extend this disciplined regime of accurate measurement to the body as well. Measuring things – as William Thomson recognised – was a way of owning them, and measuring the body was therefore a route towards its colonisation by increasingly triumphant imperial physics.

Many medical electricians were keen to display their electrical credentials, and anxious to demonstrate too their affinity with working electricians by embracing this physicist's and electrical engineer's world of precision measurement. In his *Handbook of Electro-therapeutics*, translated into English in 1883, the German medical electrician Wilhelm Erb insisted that properly understanding Ohm's law of the relationship between electrical current, potential difference and resistance was central to the proper practice of medical electricity. As Erb would have it:

> the human body is nothing more than a large conducting mass of definite resistance; and the laws controlling the distribution of large conducting masses therefore apply to it without any limitation. These are the well-known Ohm's Laws, and their accurate knowledge is of the greatest importance to the electro-therapeutist. Only by constantly bearing them in mind and utilising them in a rational manner will you be enabled to apply the current rationally and scientifically. I know of nothing in physics which is of more importance to the electro-therapeutist than an accurate knowledge of these very laws.[3]

Armand de Watteville, the physician in charge of the electrical department at St Mary's Hospital bragged:

> I still insist upon measurements of current strength as the essential conditions of a rational application of electricity in medicine. It is gratifying to me that the proposition I first made in 1878 of adopting the milliweber, now called the milliampere, as the electrotherapeutical unit, has received the sanction of the special committee of the International Congress of Electricians.[4]

Not everyone agreed. A bad-tempered little spat in the pages of the *Lancet* back in 1877 made it clear that there was more than one view on the utility of measurement in medical electricity. In a letter to the venerable journal in March of that year, de Watteville glibly pronounced on the importance of accurate measurement. 'The want of proper measuring instruments must have been deeply felt by all physicians who have anything to do with electricity,' he declared. The business needed 'a galvanometer so constructed as to give at a simple reading the actual strength of current passing through the patient's body. This reading, multiplied by the time during which the operation lasts, will give the actual quantity, literally the dose, of electricity' so that the patient could 'carry on the treatment at home with all the accuracy desirable'.[5] Not so, said Herbert Tibbits, firing off an intemperate response to de Watteville's moral and physical certainties. Tibbits was convinced that 'the dose of electricity proper to be administered in a given case cannot be measured by a galvanometer and the endeavour to do so will inevitably, sooner or later, end in disaster'. According to Tibbits what was needed was a fine touch, not a standardised armoury: 'No man is justified in electrising a patient without invariably testing electricity upon himself, if only a second at a time, and always when he either increases or decreases the power of the current.'[6] There were some very different views going around as to just what physicists' apparatus and attitudes could achieve when applied to the body.

William Henry Stone, of course, had no doubt at all that measurement was key. He had already said as much to the Society of Telegraph Engineers. As he talked to them about his aspirations to make medical electricity into a branch of physics, Stone presented the gathering with the results of his own efforts in that direction with a discussion of a series of experiments on the electrical resistance of the human body. What these experiments showed, he said, was that his fellow medical electricians needed to adopt the metrological practices and technologies of physics. Most of what they did at the moment was nothing more than 'a heterogeneous mixture of loose statements, doubtful opinions, and erroneous therapeutics'.[7] Loose language, loose theory and poor therapeutic outcomes were all bound up with lax practice. Medical electricity needed a good dose of what the physicist ordered. From his point of view, discussions in the previous year's International Electrical Conference in Paris on the choice of electrical units in electrotherapeutics

represented a step in the right direction. Stone's own preoccupation with measuring the resistance of the human body was hardly coincidental, of course. The
development of standard units of electrical resistance was an ongoing concern for
telegraph engineers and physicists alike, and a major skirmish in the war between
Britain and Germany on just this issue had been fought at the Paris Congress.
Lieutenant Colonel Charles E. Webber, the Society's president, strongly endorsed
Stone's message in his presidential address, arguing that medical men needed to
subject their 'theoretical ideas and experimental results to discussion and criticism
by professional electricians'.[8]

A few years later the same thorny question of the human body's electrical resistance was raised once more, at a meeting of the Institution of Electrical Engineers
(formerly the Society of Telegraph Engineers and Electricians). The lecture this time
was delivered by Henry Newman Lawrence and Arthur Harries on 'Alternating
v. Continuous currents in relation to the human body'. The two experimenters
emphasised that their particular position at the Institute of Medical Electricity
provided them with the unparalleled opportunity of experimenting directly on
human bodies, rather than having to extrapolate the effects of electricity on the
usual menagerie of rabbits, dogs and even horses, as did their rivals in the field. They
had subjected the same group of ten people to alternating and continuous currents
respectively, with their skin dry, moistened with distilled water and moistened with
salt water. The results varied dramatically. In the case of continuous current, for
example, the highest recorded resistance with dry skin was at 50,000 ohms, whilst
the lowest recorded resistance with skin moistened with salt water was ten times
less. One of the things they wanted to do with this data was assess the sensitivity of
the human body to electric shocks from different sources of current – a sore point
given ongoing debates about the competing virtues of direct and alternating current systems of electrical power distribution.[9]

Right at the heart of these discussions and debates about electrically measuring the human body, was the question of just what kind of electrical equipment
the body itself was. It is important to recognise just what this kind of talk took for
granted. For the most enthusiastic proponents of metrology applied to the body, the
issue was not whether the body was a piece of electrical apparatus, but what kind
of electrical apparatus the body was. Delivering the Lumleian Lectures to the Royal
College of Physicians in 1886, Stone speculated that as far as its electrical characteristics went, the human body behaved something like a piece of the Atlantic
telegraph cable. This was a useful comparison to be able to draw, of course, since
by the 1880s the Atlantic cable was probably the most comprehensively understood
piece of electrical equipment in existence – the flow of commercial information
back and forth between Old and New Worlds depended on knowing its idiosyn-
crasies thoroughly. The eminent electrical engineer and enthusiastic investigator
of spiritualism, Latimer Clark, suggested that the body's resistance could best be
understood on the basis that 'the human frame was exactly like a galvanometer'.[10]

What this kind of speculation meant, of course, was that as far as these ambitious physicists and electrical engineers were concerned, the human body really was something like their electrical components and could be fitted into their electrical networks just as neatly and usefully.

Discussions about the possibility of using electricity as a new and scientific means of executing criminals – a sure and humanitarian alternative to the chancy uncertainties of the rope – had been aired well before 1890. Promoters of electrical executions averred that their method was 'decorous, involving no brutal or barbarous intervention of the executioner; it is humane, involving no prolonged agony for the condemned'. Their logic was straightforward enough: 'In death by electric current life is undoubtedly extinct before the afferent nerves can carry to the brain intelligence of the stroke that has fallen. It follows that death by such means – death by electricity – must be painless.'[11] Different methods of turning the theory of electrical execution into practice were eagerly canvassed. One enthusiast suggested 'a small wooden house, like a sentry-box or a watchman's hut', with one electrode connected to a plate on the floor and the other hanging down from the roof. The prisoner would be led in barefoot (to ensure a good connection), secured and left to expire at the touch of a button; 'and even before the bystanders have consciousness of the act of pressing the button all is over'.[12] Most advocates eventually settled on a chair as the best method for securing the subject. The Medico-Legal Society of New York set up a committee to investigate the matter, who duly recommended that:

> the convict should be placed in a recumbent or semi-recumbent attitude, and properly secured and that 'the current used should have a potential of between 1,000 and 1,500 volts, according to the resistance of the criminal, and should be an alternating current with alterations of not fewer than 300 per second; and that such a current should be allowed to pass for from 15 to 30 seconds'.[13]

There was more going on behind these deliberations than immediately meets the eye – particularly as far as the choice of current was concerned. A vicious, no-holds-barred commercial battle was taking place between the Westinghouse and Edison electrical companies. The rivals advocated very different methods of generating and transmitting electricity. Edison wanted a system of direct current transmission at relatively low voltages, using a network of small power stations on every city block. Westinghouse's system used alternating current to transmit electricity at far higher voltages over greater distances from large central stations. This battle of the systems soon spilled over into debates about electrical killing. Edison and his supporters pushed hard in favour of the process – with Westinghouse's alternating current as the chosen instrument of death. Edison even suggested that to 'westinghouse', as in 'following his conviction the criminal was duly taken away to be westinghoused', would make a good name for the grisly business. Edison's hired hands set to work, demonstrating that alternating current really was the current that killed. Harold

T. Brown, Arthur Kennelly and Edward Tatum carried out extensive public experiments, submitting rabbits, dogs, oxen and horses to the current to show that dose
for dose, alternating current (a.c.) packed a stronger and more lethal punch than
direct current (d.c.).[14]

When the New York state legislature passed their bill making electricity the
state's chosen method of execution early in 1888, they had left the details of the
process open. It had taken much lobbying by Edison and his supporters to make
sure that alternating current was the system favoured. All that remained then was
to select a victim. The bill became law on 1 January 1889, but it was May before
the trial of William Kemmler provided an experimental subject for the first legal
killing with electricity. Kemmler had been convicted of murdering his common
law wife, Tillie Ziegler, following a domestic argument. Kemmler's conviction set
off a flurry of frantic legal activity. Kemmler's lawyers were doing their damnedest
to overturn the sentence of death by electricity on the grounds that it constituted
cruel and unusual punishment – and therefore was unconstitutional. Rumour had
it that William Bourke Cockran, Kemmler's appeals attorney, was being paid for by
George Westinghouse, desperate to keep his system from being sullied by association with electric death. Cockran indignantly denied the charge, though that did
little to prevent the New York Times from reporting it as fact. His efforts were, in any
case, to no avail. Defeated both in the New York Supreme Court and the Court of
Appeals, by the summer of 1890 Cockran – and Kemmler – had run out of options.

Even if he had spent the whole of 1890 trapped inside a Faraday cage, our
Mr Jeffery can hardly have failed to notice the furore that followed William
Kemmler's electric death on 6 August that year. Despite all assurances of a quick and
painless passing, Kemmler's execution was a botched, messy and disgusting affair.
The New York Times reported events in the run-up to the execution on a daily basis.
They speculated about the prison warden's state of mind, wondering if he really
wanted the first electrical execution on his conscience. On 5 August they reported
that invitations had been sent out to those lucky few who were to be privileged to
attend the deadly experiment. On 6 August they reported 'Kemmler's last night',
his daily routine and his desire to die 'like a man'. He promised the warden, they
reported, that 'I won't break down if you don't'.[15] The next day, with the execution
over, however, their tone had changed. 'Far worse than hanging', their headline
declared, as they condemned the entire business as a 'sacrifice to the whims and
theories of the coterie of cranks and politicians who induced the Legislature of this
State to pass a law supplanting hanging by electrical execution'. Despite the promises that this would be a quick and painless killing, it had taken two jolts of current
to dispatch Kemmler. The New York Times reporter described the grisly scene as the
current surged through Kemmler for the second time:

the witnesses were so horrified by the ghastly sight they could not take their eyes
off it … Blood began to appear on the face of the wretch in the chair. It stood

on his face like sweat … An awful odor began to permeate the death chamber, and then, as though to cap the climax of this fearful sight, it was seen that the hair under and around the electrode at the base of the spine was singeing. The stench was unbearable.[16]

The *New York Times* were in no doubt that this first electrical execution should also be the last. 'All the men eminent in science and in medicine' who had been invited as witnesses, they declared, 'almost unanimously say that this single experiment warrants the prompt repeal of the law'. In London, *The Times* also reported that 'witnesses say that the sight in the execution-room was sickening, and the general belief is that the system is a failure, and that execution by electricity as performed at Auburn will be abandoned'.[17] In an accompanying editorial the newspaper dryly speculated that the whole enterprise 'does not seem to have been attended by such a measure of success as will be likely to encourage a repetition of the experiment'. They noted that five more victims were already lined up to meet the same end as Kemmler and hoped that 'the more masculine minds of America will feel it to be intolerable that the fate of these men should give occasion for any repetition of such scenes as those which have just been enacted'.[18] These were strong words from *The Times* and must have left readers in little doubt concerning editorial revulsion. The revulsion was widely – but certainly not universally – shared on both sides of the Atlantic.

The doughty old *Lancet* led the furious charge of medical opposition as electrocutions continued despite protestations, though what worried them most was the disgrace the entire proceedings had heaped upon the heads of the medical profession. 'Could anything more uncertain or unsatisfactory be conceived,' their editor thundered, 'anything more painful be recorded on the face of the history of modern medicine? The best minds will stand astounded that any medical man or men could be found ready to dip their hands into any part of such fearful work.'[19] The defence of scientific and humanitarian precision was no defence at all as far as the *Lancet* was concerned. As electrocution followed electrocution in the months and years after Kemmler's killing, the *Lancet's* editors thought the evidence stacking up was far from unambiguous in electrocution's favour. And even if the process really was everything it was made out to be, so what? 'The death in this case is reported to have been painless,' the editorial snorted as news of the latest electrocution crossed the Atlantic by telegraph, 'and we have no doubt on that point. It was as painless as death from a blow from a club. But what a display of scientific refinement in such a deed!'[20]

The *Electrical Review*, on the other hand, exploded in outrage at this slur on the professional integrity, skill and knowledge of fellow electricians. The theory and the morality of electrocution were just fine, and the accumulation of practical experience as the death toll mounted would iron out any problems there were with the actual execution of the process. The only problem as far as they were concerned was lack of practice; that, and the bad attitude of people like the editors of the *Lancet*. The *Electrical Review* laid into them with real and revealing venom; their opposition was symptomatic of 'an attitude quite inconsistent with that which becomes a great organ of medical and surgical science' and seemed to 'resemble the hysterical ravings of an antivivisectionist'.[21] A few years later they were even more outraged with the *Lancet*'s attitude. 'We are compelled to outrage the English language and to coin a new word to express the utterly hopeless condition of our contemporary,' they ranted. 'Over the subject of electrocution as practised in the United States, it has sunk into the most maudlin state of old-womandom that it is conceivable for any leading organ to fall into.'[22] Just what had the *Lancet* done? They had raised the horrifying scenario that electrocution might not kill after all, but simply render its victims comatose – meaning that they only died in when cut open by doctors on the autopsy table.

The editors of the *Lancet* were not the only ones to worry about that dreadful possibility. Even before Kemmler's execution questions were being asked about just how electricity caused life to end. In 1889 the *Electrician* ventured that 'humanitarians and members of the medical profession' (an interesting distinction) should:

> first study the human body as an electrical apparatus, and a very delicate one; let them follow the current in each of its numerous paths inside the body, and ascertain what its effect is on each organ, each membrane, each nerve, &c., and then, if they think well, let them arrange to execute by electricity.[23]

In the United States, as those two Edisonian acolytes, Harold P. Brown and Edward Tatum, experimented on various animals to establish that a.c. rather than d.c. really was the 'current that kills'. They were interested in just how it killed, too. Tatum's view was that 'the essential field of action lies within the substance of the heart itself' and that 'electrical currents can not finally arrest respiration without simultaneous or still earlier arrest of the heart; and that there are no changes observable in the substance or functions of muscles, nerves, or blood, which can in any sense be the cause of sudden death'.[24] In other words, electricity killed by stopping the heart – an unambiguous and irreversible ending.

That discussion of 'Alternating v. Continuous Currents in Relation to the Human Body', which Henry Newman Lawrence and Arthur Harries delivered to the Institution of Electrical Engineers at the end of March 1890, certainly needs to be appreciated in this broader context of the vicious battle of the systems raging between Edison and Westinghouse on the other side of the Atlantic. As well as the rather more immediate context of Kemmler's approaching execution. What the two electricians presented there, after all, were the details of experiments on human subjects that had been designed to reveal comparative measurements of resistance to alternating and continuous currents. They were also trying to provide an objective measurement of what might to many seem a very subjective experience: the relative degree of discomfort that was produced by alternating and continuous currents respectively. The lively discussion that followed the presentation – during which most of the respondents rejected the presenters' conclusions to some extent or other – mainly focused on the reliability and accuracy of the instruments, whether the experimenters were really measuring the right thing and on a great deal of anecdotal evidence. Member after member lined up to reminisce about great shocks that they had survived, with the irascible Welshman William Henry Preece, the Post Office's superintendent of telegraphs, claiming the laurels in that respect with a highly respectable 2,000 volts.

Newman Lawrence and Harries repeated their performance at that year's BAAS meeting in Leeds. Preece was there again and just as anxious to highlight the vari ability of bodily response to electricity, suggesting that more people died of fright on receiving a shock than from any actual physiological effect. He thought that 'it was startling to find that alternating currents were so much more painful than con-tinuous currents, and also startling to find that these sensations of discomfort could be produced with such ridiculously little currents'. Preece was sure that he and other experienced old hands knew better what was really going on and what the human frame (or their robust physiologies, at any rate) could tolerate. After all, he boasted, 'in his career, he must have received a million shocks, or perhaps more'.[25] The *Electrical Review* thought that Preece's antics on this and other occasions were risible, calling him 'an incorrigible wag' on the matter:

His remarks on the subject of danger from electric currents were so severely criti-cised at the time of the British Association Meeting that we should have given him credit, when another occasion arose to display his profound contempt for shocks, for more moderation, but here we find him going further and further out of his depth.[26]

By the time of the BAAS Leeds meeting, the New York execution had taken place. Even Preece suggested that 'they had a great deal more to learn before they dared to use electricity for the purpose of execution'. Harries agreed, saying the 'recent affair in America was, in his opinion, bungled in a fearful manner'.[27]

The physiology of death by electric shock was a regular item in the correspond-ence pages of the *Lancet* for much of the 1890s. Following on from Newman Lawrence's and Harries' original investigations, the *Electrical Review* complained bitterly about 'the lamentable lack of electrical knowledge displayed by some of the medical correspondents'.[28] Some medics also thought the electricians woefully ignorant, pointing out that the 'difference which Dr. Harries and Mr. Lawrence have found in the resistance of the body, according to whether alternating or constant currents are used in the measurement is an interesting consequence and illustra-tion of facts well known to physiologists'.[29] The flamboyant Nikola Tesla's visit to London in February 1892, and his spectacular performances with high tension,

high frequency currents before the assembled ranks of the Institute of Electrical
Engineers meeting at the Royal Institution, caused much excited speculation too.
Tesla's lecture was a bravura performance that made graphic the ambiguities of
bodily electricity. 'Here is a simple glass tube from which the air has been partially
exhausted', went Tesla's showman's patter:

> I take hold of it; I bring my body into contact with a wire conveying alternating
> currents of high potential, and the tube in my hand is brilliantly lighted. In whatever
> position I may put it, wherever I move it in space, as far as I can reach, its soft pleas-
> ing light persists with undiminished brightness.

Tesla proceeded to place himself in the same circuit as light bulbs, which would be
'filled with magnificent colors [sic] of phosphorescent light', or with a large induction coil so that 'streams of light break forth from its distant end, which is set in
violent vibration'.

In Britain, researchers also turned to the work of French physiologist Arsène
d'Arsonval on electrophysiology, which indicated that in many cases electrical
deaths were caused by asphyxiation.[30] The Frenchman was, among other things,
famous for his work on high-tension electrical currents, demonstrating that they
could be conducted safely through the human body – and even using them to power
an electric light held in his hands. Summarising this work, the electrotherapist W.S.
Hedley suggested that there were two roads to electrical death: electricity could
kill either by 'actual lesion of tissues' or by 'arrested respiration causing asphyxia'.[31]
In the second of these cases, resuscitation was possible. Good news as this might be
for the potential victims of accidental electric shocks, it was this conclusion that
led medical opponents of electrocution to raise the nightmare scenario that 'more
murderers have paid the penalty for their misdeeds on the dissecting table than
in the electric chair'.[32] In 1895 the physician Henry Lewis Jones – electrotherapist and chief medical officer at St Bartholomew's Hospital electrical department
– challenged d'Arsonval's conclusion that respiratory failure was the primary cause
underlying death from electric shock. The results of autopsies did not fit this model,
he pointed out. For example, the 'heart is often noted to be relaxed and empty,
and this is not in accord with the theory of death from asphyxia'.[33] Assuming that
shocks were always recoverable from might lead to complacency, he warned. He
described a number of experiments on cats which he reckoned 'showed conclusively that death was caused by failure of the heart' and that 'the action is upon the
heart muscle rather than upon its nervous mechanisms'.[34] A few years later in 1898,
another contribution to the British Medical Journal put forward similar arguments
and came to the same conclusion that death was caused by cardiac rather than
respiratory failure, drawing on the testimony of workmen 'that when one of their
mates has been fatally injured by coming into contact with highly charged metal,
he has generally breathed a few times although to all appearances dead'.[35]

Hedley had argued that 'in most cases of accidental contact with electric-light wires the condition is generally one of suspended animation – of only apparent death'.[36] As far as he was concerned, like d'Arsonval and the anti-electrocutionists (and William Henry Preece for that matter), it was 'not always so easy as might be supposed to kill an animal by electricity'. He offered, nonetheless, a chilling account of how electricity could and did kill in the end:

Entering the skull, a current diffuses itself through the brain, sweeping away the faculties of sensation, intellect, and will; concentrating itself, it makes for the foramen magnum, smiting in its path that vast collection of conducting fibres and nerve centres – the medulla oblongata. The delicate mechanism by which vital functions (respiratory, vaso motor, cardio-inhibitory &c.) are governed is thus destroyed, and it is on the balanced relationship of such functions that life depends.[37]

More broadly, Hedley thought of the human body as anchored in a sea of ether. Any conductor, 'whether it be a wire or a living body, only guides the energy and concentrates it for useful work'. So the body was just:

an appliance capable of utilising in a variety of ways energy transmitted by the ether. It resembles an ordinary metallic conductor in that it guides the effects transmitted by the ether contained in the dielectric, but it differs from the wire in the fact that the guidance is probably effected by that peculiar form of conduction known as electrolytic.

In which case ether vibrations might be imagined as 'impulses communicated, motion added, to the already pulsating atoms of a vital structure'. Parts of the body's mechanism could be visualised as 'receivers', 'syntonised for the reception of similarly vibrating etherial impulses radiating from some given source'.[38]

In a way that managed to be simultaneously practical and metaphysical, at stake in these kinds of debates was the status of the ultimate barrier between death and life. What is particularly striking is that one of the few things all the protagonists appeared to agree on was the permeability of that final frontier. Death in the 1890s, on this account, appears to have been remarkably negotiable. For much of the nineteenth century, 'electricity is life' had been the common slogan. After 6 August 1890, that must have sounded hollow, to say the least. At the beginning of the century, Aldini's experiments had seemed to offer the possibility of electrical revivification. During the 1870s, however, Benjamin Ward Richardson cast doubt on this reading of electricity's relationship with vitality. According to Richardson, galvanic stimulation of the kind Aldini or Ure had attempted could only 'create a semblance of life without a reality'. At best, electricity was doing something that was 'unpardonable in its horror'; it was 'restoring a momentary intelligence which we cannot sustain' and 'enabling the prostate body to look into life only to sink back into oblivion'.[39] In 1869 Richardson had publicly experimented on the electrical properties of animal matter in a series of flamboyant lectures at the Royal Polytechnic Institution, using their newly purchased 'monster coil'.

By the end of the 1890s, the veteran electrotherapist Julius Althaus was again arguing that electricity was the key to life. Electricity judiciously applied to the brain offered rejuvenation. Old age was caused by 'degeneration of the neuron' and could be delayed – or even reversed – by providing additional nutrition to the neuron by 'careful and appropriate use of the constant galvanic current'. Althaus suggested that he had discovered that:

> a cautious use of the constant current to the brain, and more especially to the vaso-motor centre in the bulb, greatly to retard the progress of the arterio-sclerosis and the involution of the central neuron ... The old man takes fresh interest in the affairs of daily life, he resumes his work with some amount of vigour, he has a more erect attitude, he walks and stands better, and he has a quicker digestion and a healthier sleep.[40]

The electric current worked both ways if Althaus was to be believed – it could kill and, not just cure, but give the dying a new lease on life. Electricity and its practitioners straddled late Victorian unease about mortality, and concerns about the ambiguous status of the line between the quick and the dead.

If Mr Jeffery read his Arthur Conan Doyle he would have found the author of *Sherlock Holmes* toying with that very conundrum in a short story published in 1894. In Conan Doyle's tale, the good citizens of Los Amigos, somewhere in the Wild West of America, reacted to news of the use of electrocution as a means of execution in typical booster fashion. They would outdo those damn Yankees by building bigger and better: the 'Western Engineers raised their eyebrows when they read of the puny shocks by which these men had perished, and they vowed in Los Amigos that when an irreclaimable came their way he should be dealt with handsomely and have the run of all the big dynamos'.[41] When they duly got their man they gave him the full 6,000 volts their largest generators could manage; and the result? 'There he sat, his eyes still shining, his skin radiant with the glow of perfect health, but with a scalp as bald as a Dutch cheese, and a chin without so much as a trace of down.' What had they done? They had 'increased this man's vitality until he can defy death for centuries ... it will take the wear of hundreds of years to exhaust the enormous nervous energy with which you have drenched him. Electricity is life and you have charged him with it to the utmost.'[42]

Conan Doyle clearly knew how to have fun playing with electricity's ambiguities. There was a serious point there, nevertheless. Electricity during the 1890s was a commodity up for grabs. It seemed to have the potential to be and do all sorts of different things to different people – some of them mutually contradictory. Competing groups of people with their own various claims to being the voices of

authority were squabbling over the electrical body. Measuring the body's characteristics, surrounding it with electrical instruments, subjecting it to standardised
currents – these were all ways of trying to turn the electrical body into cultural
property. No wonder that someone like Mr Jeffery might end up more confused
than enlightened. Maybe that was why he turned to Cornelius Bennett Harness
and his electropathic belts instead of going to see the likes of Armand de Watteville.
His message might be of rather dubious provenance, but at least it was fairly plain
what he was offering. Thanks in part to our Mr Jeffery, Harness' happy-go-lucky,
unscrupulous salesman's hucksterism and the self-disciplined culture of corporeal
electrical metrology were about to enter on a collision course. An English court of
law was about to decide who really were the authorities over the life and death of
the electrical body.

SIXTEEN

HARNESS ON TRIAL

As things turned out, taking Mr Jeffery to court over the relatively paltry matter of an unpaid IOU, amounting to a mere £3 5s, would cost Cornelius Bennett Harness and the Medical Battery Company very dearly indeed. It is rather difficult to understand just why a canny operator like Harness took such a risk in the first place. Jeffery can hardly have been the first of his customers to hand over an IOU and then refuse to cough up when they discovered that the electropathic belt failed to deliver the hoped for and promised miracle cure. Was Harness' success going to his head at last? It seems clear that until he met his nemesis in the humble figure of Mr Jeffery, his business had been very successful indeed. Had the reckless gambling streak, which had fuelled his seemingly unstoppable momentum and characterised his career so far, finally turned against him? The impression one gains from observing his increasingly frantic manoeuvrings through the early years of the 1890s is certainly that he was a man bent on digging a deeper and deeper hole for himself. More seriously though, Harness' headlong fall from grace illustrates a sea change in the culture of electricity at the end of the Victorian era. He found himself trying to face down the ever more unstoppable juggernaut: a powerful and increasingly dominant culture of instrumental precision. Unsurprisingly, though maybe by no means inevitably, it rolled straight over him.

This was not, as it happens, Harness' first flirtation with using the law as a potential weapon in his ongoing battle to dominate the electrical health market. A few years previously, for example, he had had the temerity to threaten a lawsuit against no less an institution than the venerable *Lancet*. Following the Medical Battery Company's acquisition of the Zander Institute, and its lucrative trade in providing Swedish gymnastics and massage to its clients, Harness with his usual chutzpah started publishing advertisements and pamphlets listing the names of prominent doctors who had sent their patients along to the previous proprietors for consultation and treatment. Predictably enough, the indignant doctors – who included the President of the Royal College of Physicians – protested in the pages of the *Lancet* that they had never permitted their names to be exploited in such a fashion, and certainly did not want their reputations tarnished by association with the Zander Institute's new manager. Harness responded with an equally indignant letter threatening to hold the *Lancet* 'responsible in damages for your unjustifiable libels, and you will hear from our solicitor in due course'. The *Lancet* responded by publishing Harness' threat, as well as the subsequent solicitor's letter, reaffirming their view that nothing said about the Zander Institute and its machines 'when it was controlled and responsibly managed by an accomplished medical man' could be taken to apply 'when under the control of a "medical electrician" of whom we know nothing except through the medium of his advertisements'.

Harness, on that occasion, had the good sense to let matters rest. He presumably thought that Mr Jeffery would be rather more easily intimidated than the editors of the *Lancet*. Jeffery, however, was clearly in no mood to roll over and pay up. He responded instead by slapping down a counter-claim of his own against the Medical Battery Company for misrepresentation. The case of the Medical Battery Company v. Jeffery, as well as Jeffery's counter-claim, was duly heard at the Bloomsbury County Court on Tuesday 19 July, with his honour Judge Bacon presiding. Mr De Witt, as the Medical Battery Company's barrister, started the proceedings with a fulsome description of his client's reputation. The company had, he said, 'acquired a high reputation through their sales of electrical appliances for the cure of diseases. His Honour was probably aware of Harness' patents for electric belts and similar appliances'.[1] De Witt described how:

> the defendant ... came to the plaintiff's place and said he was suffering from a sprain. The receiving officer of the establishment at once introduced him to Mr Simmons, an expert in these matters in the employ of the plaintiffs. Mr Jeffery said he had a sprain and was afraid it might turn into a rupture. Mr Simmons examined him, and told him it was desirable that he should wear one of Harness' electric belts with a suspender, in order to obviate what might develop into a rupture.

As Jeffery's lawyer, Mr Lickfold, soon made clear, the case would largely hinge on just what Jeffery had said when he entered the premises and what exactly had been said and promised to him. There had been no breach of warrant, blustered De Witt, and Simmons would 'go into the box and say he never warranted that the belt would cure the defendant'. Lickfold scoffed in return that he would not dream of suggesting 'anything so absurd as that anybody warranted that the belt would cure anything'. The claim was for misrepresentation and it rested on the claim that Jeffery on entering the Medical Battery Company's premises had asked to see a 'qualified man' and had been misled into believing that Simmons was such a person. It became clear during cross-examination that Simmons, the consultant, had no medical training, having been previously employed as an Oriental furniture salesman. Simmons argued, however, that since he had studied hernia 'ever since I was nine years of age', he was more qualified than the Institute's medical men to diagnose and treat such cases on the grounds of his practical experience.[2] He claimed that there had been two qualified medical men on the premises on the

occasion of Jeffery's consultation, but that he had not called them since he did not
'think it absolutely necessary to consult them in case of hernia'.

Lickfold pressed the unfortunate Simmons hard on just what he thought he
was doing: 'Is electricity a good thing for hernia?' he queried. 'Yes, in some cases.
In the first, or incipient form of hernia, electricity would be a good thing,' came
the response from Simmons. 'What would be its effect?' interjected the judge. 'It
would act as a tonic to the muscles – to the weak parts,' averred Simmons. 'What
would be the effect of electricity?' continued Lickfold. 'I cannot say, but the belt
would give support,' Simmons admitted. 'Would the electricity heal the hernia?' he
was asked. 'No,' was the only answer. Lickfold eventually badgered Simmons into
admitting: 'I have no knowledge of electricity.' At which point he exploded, 'Then
why did you sell an electric belt and charge £5 5s for it?' Simmons could only
babble that he had 'a sufficient knowledge' of electricity's 'physiological effects'
to know that it would provide 'a tonic to the local parts' in this case. Called to
the stand and cross-examined by Harness' barrister, William George Rowntree,
Jeffery's doctor, stated that when the poor man eventually came to see him there
was 'no doubt about the nature of his suffering'. He had examined the offending
belt as well and was sure that 'if it was fitted tight on a patient suffering from rupture it would aggrevate the complaint'.

T.E. Gatehouse, one of the proprietors of the *Electrical Review*, was called to the
stand and testified that the electric belt as recommended to be used by the Medical
Battery Company could not produce any electricity, and that even in circumstances
in which the belt could produce electricity, any current would pass through the
belt itself rather than through the body. Gatehouse had come suitably prepared to
defend his assertion that the belt had 'never produced a trace of electricity' if worn
in the manner Harness recommended. 'This morning I made an experiment with
this belt, so as to produce a small quantity of electricity,' he reported. 'I took salt
water, and moistened the two parts, and then connected the copper and zince with
a galvanometer, and so I got a slight deflection of the needle of about 14 degrees.'
So there were circumstances under which Harness' belt could produce a small electrical current. However, he continued:

> I then put the belt on my legs, and on my arms, and the deflection was not in the
> slightest degree altered. That shows that when these belts are worn by any patients
> no electricity passes through the body in any way whatever, but only along the
> webbing of the belt and over the skin surface. Electrically these belts are useless.

The Honourable Judge Bacon duly concluded for Jeffery and against the Medical
Battery Company. The defendant, Judge Bacon said, had been 'taken in in the sense
that he had been persuaded to buy something utterly useless'. This in itself was not
necessarily sufficient to allow him to break his contract with the company or claim
recompense. But it was also clear, said the judge, that Simmons' standing had been

misrepresented and that Jeffery had based his decision to buy the worthless belt on the basis of this misrepresentation of the former Oriental furniture salesman as an expert. As far as electricity went, Judge Bacon was rather more equivocal. 'I will not enter into the merits of an electrical appliance which I do not understand,' he sniffed. 'Mr. Gatehouse's evidence as to the electrical volume may be right or wrong – I do not know; but this (this belt) is the thing put on the man when he was ruptured, and the man was persuaded to buy it upon the representation that he was not ruptured.' Judge Bacon was to go on to develop something of a reputation as a sceptic as far as scientific experts were concerned. Nine years later the *British Medical Journal* noted:

> that Judge Bacon, in the Bloomsbury County Court has now advanced so far as to offer to have a witness sworn in the Scotch fashion. It is true he could not forbear a sneer at a witness who had made a pretence of kissing the Book, 'That is too transparent;' said the judge, 'if you have got a fad about microbes, you should say so, and I will swear you Scotch fashion'. This offer marks a distinct advance, although it would have been more satisfactory had the judge set an example and carried out the law in a more gracious fashion. Possibly Judge Bacon is one of those people who have never seen a microbe through the microscope, and who have not taken the trouble to make themselves acquainted with the elementary facts of bacteriology. They are known to the man in the street, but the ignorance of judges with regard to matters of common knowledge is proverbial.[3]

The *Electrical Review* in reporting on the matter was predictably insistent that electricity really was at the heart of the matter, and that the public needed to place their trust in proper expertise rather than in miracle cures:

> If the much to be pitied individuals of both sexes who so readily fall into the snare set by the president of the *British Association of Medical Electricians*, and of the Medical Battery Company, would only reason with themselves, a few minutes reflection ought to convince the most foolish that the idea of a poor debilitated body generating the electricity for alleviating its own suffering is most preposterous. Indeed the stepping stone from hallucinations of this kind to the borderland of profound belief in perpetual motion can be but a narrow one.[4]

Issues of morality and decorum were at stake here as well. The *Electrical Review*, like the judge, was incensed that an unqualified man had presumed to conduct a medical examination on the body. They expressed the fear that even worse outrages might take place as well:

> [i]t is bad enough that the male sex should submit to be 'examined' by these so-called 'experts', but to think that gentle, refined, and highly sensitive women should

be contaminated by the mere touch of such 'consulting officers', is utterly abhorrent to the feelings of anybody possessed of a trace of morality or a scrap of manliness.[5]

It was not long before Harness was embroiled again, albeit indirectly, in court action to defend his reputation, his business and his claims to expertise over bodily electricity. On this occasion his opponents were the editors of the *Electrical Review* themselves. The plaintiff, however, was not Harness but our old friend Herbert Tibbits. Tibbits' accusation of libel followed an attack by the *Electrical Review* on a pamphlet he had written describing experiments showing that Harness' electropathic belts were indeed a source of electricity. The *Electrical Review*'s editors maintained that 'the writer exhibits a most incredible ignorance of electrical laws, an ignorance which utterly unfits him to speak as an authority, and which in view of the support which it affords for the bolstering up of this appliance, seems to call for an interdiction by the British Medical Association'.[6] The case was particularly serious, they argued, since Tibbits was 'supposed to be a recognised authority in the medical world'. Tibbits' situation had clearly changed a great deal since the halcyon days of the West-end Hospital for Nervous Diseases, Paralysis and Epilepsy. Tibbits had left that establishment during 1892 following a violent disagreement with the hospital's Committee of Management over financial matters. The disagreement focused on the organisation of a failed fundraising bazaar for the hospital, for which Tibbits had advanced a large sum of money and which the committee refused to reimburse following the event's failure.[7] One result of this fiasco was that, at some time during 1892, Tibbits was declared bankrupt. Under those circumstances, when approached by Harness and asked to perform a number of experiments on his electropathic belts to determine their capacity to produce electricity, Tibbits agreed to do so in return for a fee.

In his report on the experiments Tibbits carefully limited the scope to two central issues. The experiments aimed to determine whether the belts actually generated electricity, and if so, whether the electricity so produced could penetrate the skin so as to act on the underlying muscles, nerves and organs.[8] The belts' capacity to produce electricity was first tested by slightly moistening them and then by wearing them according to instructions so that they were set in action by natural perspiration. Under both circumstances a reading was obtained on a galvanometer. With the belt worn on the body and one pole connected to a Kelvin reflecting galvanometer, a platinum needle insulated to within one-sixteenth of an inch of the tip was inserted through the skin, and also connected to the galvanometer. The instrument registered a reading of several degrees, suggesting, according to Tibbits, that electricity from the belt did indeed penetrate the skin. To test the matter further, Tibbits conducted another experiment on the body of a recently slaughtered and skinned rabbit, again showing that electricity from the belt actually entered the body rather than passing along the surface. This was the crucial technical point of the case. Only if electricity could be shown to pass actually through the body could it be argued that the electropathic belt was of therapeutic value. Tibbits' and Harness' reputations rested, therefore, on establishing that claim.

Tibbits asserted that 'these experiments prove conclusively that a current of electricity penetrated the skin and influenced the subjacent tissues'. He admitted that:

> the dosage of electricity has not yet been scientifically determined, and the question with regard dosage of these weak and long-continued currents must stand the test of experience; but I claim to have settled that, as above stated, currents do enter and circulate in the body when carefully applied by means of such efficient apparatus.[9]

He went further to say that:

> I have tested many advertised appliances, and without entering into a long dissertation upon the relative efficiency of these, I have no hesitation in saying that regarded from a strictly scientific point of view, the Electropathic Appliances manufactured by the Electropathic & Zander Institute, who are manufacturers of Mr. C.B. Harness' [sic] various patents, 52, Oxford Street, London, are preferable to any others ... These appliances are scientifically constructed, perfectly manufactured, and admit of being placed upon and kept in exact and close contact with the skin, or with any part of the body.[10]

In short, Tibbits' pamphlet was nothing less than a comprehensive recommendation of Harness' products. Given Tibbits' reputation as a spokesman on medical electricity, his testimonial had the potential to transform Harness' place on the cultural map. He could be vindicated as a legitimate purveyor of medically and scientifically sanctioned products.

When the case appeared before the Queen's Bench Division of the High Court on 15 February 1893, Sir Richard Webster QC, acting for Tibbits, sought to establish that his client had, throughout, acted in good faith, that the experiments he had carried out on Harness' appliances were reliable and conclusive, and that the inspections Tibbits had carried out of Harness' premises were thorough and entirely justified the statements made in the pamphlet. The editorial remarks made by the *Electrical Review* were tantamount to an accusation that Tibbits was guilty of imposture and was 'an abettor of a gross fraud' and were therefore a 'gross and slanderous libel'.[11] The defence, on the other hand, were anxious to protect their clients' reputation as electricians and their right as men of science to criticise others.[12] They insisted that no direct attack upon Tibbits was implied in the offending article but that Harness and his business had been the sole targets. They contended that if any libel action had any validity it would be an action originating from Harness rather than Tibbits. As their cross-examination made clear, the defence was anxious to turn the trial to hinge on the reliability of Harness' electropathic belts. Their examination of Tibbits was designed to ascertain precisely what his views were concerning the belts' actions. They aimed to show that their remarks were justified by producing witnesses to testify that the experiments carried out by Tibbits did not warrant the conclusions he had drawn and that any person conversant with the principles of electrical science would not have drawn such conclusions.

To this end the defence had gathered a number of eminent witnesses. Whilst all witnesses conceded that under certain circumstances electricity would be produced by the belts, they were mainly concerned to dispute Tibbits assertion that his experiments showed that the electricity travelled subcutaneously. All argued that the electricity in that experiment had been due to the reaction between one of the poles of the belt and the platinum needle itself rather than a reaction between the two poles. James Swinburne, an electrical engineer, claimed that 'the statements made by Dr. Tibbits showed he had hardly the most elementary acquaintance with electricity and was therefore incompetent to deal with the subject he wrote about'.[13] Silvanus P. Thompson, author of electrical textbooks and popularisations and Professor of Experimental Physics at University College Bristol, suggested that 'This dear old doctor does not know what he is talking about'.[14] Armand de Watteville and H. Lewis Jones, both prominent medical electricians and authors of textbooks on medical electricity, concurred that any current produced by the belts would be therapeutically worthless. The defence's star witness was Lord Kelvin, then President of the Royal Society and probably the most eminent physicist of the day.[15] Kelvin duly pronounced it as his opinion that the belts could produce no more than an infinitesimal quantity of electricity.

Tibbits in his evidence was insistent that his experiments showed that electricity did indeed pass through the body from the electric belt. He agreed with the defence witnesses that this was the crucial technical issue: 'the important question for the public is whether or not they get any electricity at all, quite independent of dosage, as to which, in the present state of electrical knowledge, nothing has yet been decided.'[16] Several other witnesses concurred with Tibbits, including Arthur Harries formerly of the Institute of Medical Electricity, that electricity generated from the belt would pass through the body. Tibbits' reputation, as well as Harness', rested on the proper interpretation of the swing of a galvanometer needle. Tibbits' claims concerning the significance of his experimental findings and his reputation as a medical elecrician were, however, insufficient to withstand the counter-offensive mounted by some of the biggest guns in British electricity. The jury eventually needed only a few minutes to conclude that the defendants were innocent of the charge of libel. Far from the trial having succeeded in associating Harness with Tibbits' scientific credibility, Tibbits had emerged from the proceedings tarred with the same brush as Harness.

It was not long, nevertheless, before the *Electrical Review*'s editors were back in court. On this occasion they were the plaintiffs and the defendant was Harness. Harness had attempted to flex his financial muscles in response to the *Electrical Review*'s continued campaign against him by circulating newsagents with a letter to the effect that the *Electrical Review* contained malicious libels against him and that if they sold it they too would be subject to libel action. The notice circulated by Harness' solicitor warned its recipients that having been:

> consulted by my clients, the Medical Battery Company (Limited), the proprietors of Harness's Electropathic Belts and also the Electropathic and Zander Institute, of 52, Oxford Street, W., with reference to some malicious libels published in the *Electrical Review* and a paper called *Science Siftings*, I beg to give notice that if these or any further publications containing defamatory articles or paragraphs are sold or circulated by your firm, my clients will hold you responsible. In the same way you will be held responsible for the exhibition of any libellous bills or placards.[17]

Beneath the dry legalese of the letter lurked a naked threat to the *Electrical Review*'s future and its proprietors reacted accordingly. They counter-attacked with an application for an injunction to prevent the further circulation of the letter and claimed damages for the 'malicious libel' that they alleged the letter's contents perpetrated against them.

The case was heard before the Lord Chief Justice, none other than John Duke Coleridge, first Baron Coleridge and great-nephew of the poet and galvanic enthusiast Samuel Taylor Coleridge. One cannot help but wonder if the judge knew of his great-uncle's fascination Two years earlier Coleridge had been the judge presiding at the infamous gambling case of Gordon-Cumming v. Wilson in which the Prince of Wales had been forced to testify. Representing the *Electrical Review* were the medically trained Robert Finlay, who would be appointed Solicitor-General by Lord Salisbury a couple of years later, and Sir John Lawson Walton. Harness had secured the services of Sir Edward Clarke who, ironically, would be Salisbury's first choice for Solicitor-General before eventually settling for Finlay. Both sides, then, were obviously prepared to bring out the big guns. Clarke's strategy was clearly going to be one of damage limitation. He (or rather his assistant, since as things turned out he was detained at the Old Bailey at the time of the trial) submitted a defence of privilege, arguing that the circular had been sent around by his client for his own protection and with a view to preparing for his own action against the plaintiffs. What they wanted to avoid at all costs was Finlay, on behalf of his clients, turning it into a trial of Harness' electropathic belts. Which, of course, was precisely what Robert Finlay proceeded to try to do.

In his opening address, Finlay outlined the background to the case, summarising Harness' action against Jeffery and its outcome, and quoting at length from the *Electrical Review*'s account of the trial and their later editorials – which were the matter of Harness' own anticipated charge of libel against the journal. Germaine, Clarke's assistant, tried to interrupt, arguing that all this material was irrelevant to the case being tried that day. Finlay counter-argued that the details were absolutely to the point since they dealt with the defence's claim of privilege. In short, Harness could not have thought the *Electrical Review*'s editorials were libellous (and could not therefore claim privilege) because he knew that they were in substance true. When it became apparent after much legal debate that Chief Justice Coleridge was disposed to accept Finlay's arguments and therefore prepared to hear the mass of evidence that the *Review* had amassed to substantiate its claims, Germaine for the defence admitted the libel without further argument and the jury duly awarded damages of £1,000 – a not inconsiderable sum in the 1890s.[18] An editorial in the *Electrical Review* shortly afterwards made clear the extent of the evidence and witnesses who had been amassed against Harness. Indeed, they could list the names of some of the country's most eminent electrical engineers and medical electricians as having been prepared to be witnesses for their case, or as having assisted in conducting experiments on Harness' belts.[19]

In the aftermath of these events the popular campaigning London newspaper, the *Pall Mall Gazette*, decided that the time had come to deliver the *coup de grâce*. Their attack was at least partially based on some of the evidence amassed by the *Electrical Review* in its defence. Clearly that journal's editors, denied their day in court by the opposition's capitulation, were out to give their findings the greatest possible publicity too. Over six prominent front-page articles the newspaper systematically dissected and ridiculed Harness' background and qualifications, his scientific credentials and his business practices. And so in October 1893, whilst their readers were being regaled by the latest news of Cecil Rhodes' annexation of Matabeleland and of the indefatigable Dr Jameson's campaigns against the Matabele king Lobengula, they could also read all about another example of flagrant skulduggery that had happened far closer to home. In what they styled as 'one of the fiercest attacks ever made in the English press upon an individual', the *Gazette*'s attack was designed to expose Harness and his company as quacks and charlatans, and the medical electrical devices they manufactured and sold as worthless.[20] The *Pall Mall Gazette*'s editors argued that the imposition made on the credulousness of the paying public by Harness and his panoply of electropathic belts and corsets merited their forceful intervention. They more than hinted that some of their competitors, who profited from Harness' extensive advertising in their pages, were unwilling to expose his activities as the sham they perfectly knew that they were.

The *Pall Mall Gazette* was already adept at this kind of exposure. Some years earlier in 1885, under the previous editor W.T. Stead, the newspaper had played a key role in the white slavery scandal that had rocked London. In a series of articles in the *Pall Mall Gazette* on 'The Maiden Tribute of Modern Babylon', Stead exposed to his avid readership the trade in young, working-class virgins for wealthy, upper-class roués that he claimed flourished in London's underworld. Stead had posed as a buyer and successfully purchased a 13-year-old girl for £5 as evidence of the trade – ending up in prison himself for abduction and indecent assault.[21] A fervent supporter of radical politics and women's rights, Stead used the story to highlight sexual and economic exploitation. The same kind of trend can be discerned in the Harness exposé, though it took place after Stead had ceased to be editor. A ruthless charlatan – the equivalent of the 'vile aristocrats' of the 'Maiden Tribute' – was preying on the innocent, underprivileged and helpless. The 1880s and 1890s were in any case a period when both male and female sexuality was in the process of being redefined in the face of increasing calls for female emancipation and sexual freedom. Harness, with his promises of cures from sexual dysfunctions, could be portrayed, as he was by the *Pall Mall Gazette* and other publications, as an exploiter of people's sexual fears in the face of social uncertainty.

The *Pall Mall Gazette* was out to prove that Harness and the Medical Battery Company were 'fattening on a system of fraud and imposture which is absolutely unequalled in the annals of swindling'.[22] Harness, they claimed, was 'a common, illiterate, and unscrupulous charlatan' who issued 'disgusting pamphlets, teeming with lies to catch both men and women'. Those members of the public unfortunate, desperate or gullible enough to be drawn by his activities were 'hoaxed by bogus experiments and fraudulent representations ... his whole object is to get money out of them ... he preys on the ignorant poor whom any other rogue would scorn to rob, and resorts to trickery of the meanest and basest sort to make them pay'. The electropathic belt itself was 'a swindling appliance, without any electrical virtue whatever ... sold for more guineas than it is worth shillings'.[23] By the end of the series and the voluminous correspondence from a range of Harness' disgruntled former clients that it attracted, little can have been left of his or his electropathic belt's reputation. Of the letters published in the *Pall Mall Gazette*, only one clearly defended Harness and that on the rather weak grounds that his practices were no different from those of any other businessman.

'A Man of Business' expressed himself:

very much surprised at your onslaught on the Medical Battery Company ... As a man of the *world* you know that it (the world) is composed mostly of fools, and that if it is not Mr. Harness who gulls them somebody else will; and personally, though agreeing with your article as to the wholesale sham, yet I admire the cleverness of the man in making and sustaining such a business.

This particular correspondent was not entirely a disinterested customer, however. He roundly told the *Pall Mall Gazette* that:

the destruction of that concern means the loss of nearly £300 to me, and would, perhaps, be my ruin; so that when a company has been working for some twelve years you will see that with a view to a good investment for saving complete outsiders put capital into concerns even of this character, though innocent of any sham (I don't say fraud) that the profits are made upon.[24]

In this case the 'Man of Business' had certainly chosen the wrong horse to back. Within a year, the Medical Battery Company was in the hands of the receiver, although some efforts seem to have been made to revive it later in the decade. Similarly, Herbert Tibbits' reputation was in tatters. He was expelled from membership of the Royal College of Physicians and struck off the Medical Register. He had been transformed from being an eminent expert on electrical medical matters to the disreputable associate of a quack.

So Mr Jeffery had his revenge and Harness his comeuppance. There is more to the story than a simple morality tale, however. Harness' well-deserved downfall was not in any sense simply the inevitable outcome of his quackery catching up with him at last. It was symptomatic of a change in the sensibility of electricity during the final decade of the Victorian age. There was a widening chasm between the flamboyant, radical and sensational bodily electricity that went along with the culture of electrical display on the one hand, and the cool electrical discipline of the laboratory on the other. The showmen were losing their battle for supremacy over the electrical body and the new masters of metrology were winning. In the end, it all boiled down to authority. Who were people like Mr Jeffery to believe when it came to making sense of the mechanism of the body? The fate of Harness' electropathic belts, their inventor and others like Tibbits who found themselves on the losing side in this imbroglio, demonstrate just how fragile a thing authority – and scientific authority in particular – was still during the 1890s. It shows how unsure the Victorians still were about where to turn for comfort about their bodies, their selves and the world in which they found themselves. People were in the process of finding new ways to live and were groping for new ways of thinking about themselves – or rather for new people to tell them what to think about themselves – as they did so.

PART V

BACK TO THE FUTURE

SEVENTEEN

BACK TO THE FUTURE

Looking back on the early years of the 1900s, after a long and extremely successful medical career, the psychiatrist James Crichton-Browne jotted down some musings about the nature of electricity. 'There are many forces in Nature which we have not the requisite senses to perceive,' he mused, in one such jotting. 'For instance, electricity is a universal element which might well have escaped cognisance but for its occasional concentrations and disturbances, making vivid appeal to two senses, when we have lightning and thunder.' Nevertheless, he thought:

it is worthwhile to note that though we have no organ differentiated to perceive electricity, as the eye and the ear perceive light and sound (an organ which would, by the way, be useless unless we had also a power of self-insulation on the approach of danger) we have a very general physical perception of electrical changes.

Musing again in similar vein, Crichton-Browne speculated about how we:

used to try to explain electricity in terms of matter; we now try to explain matter in terms of electricity. May not electricity be yet resolved into a manifestation of a far more subtle psychical energy that is and perhaps ever will be beyond the reach of physical research, but in which we live, move and have our being?[1]

It is not surprising, perhaps, that in the 1930s someone like Crichton-Browne looking back at an illustrious career dealing with Victorian nervousness, would have had electricity on his mind. He was quite right, though, in identifying the persistent link between electricity and Victorian bodies that ran through the previous century. Many of his contemporaries clearly agreed that there was something quite visceral about electricity. For most people throughout the Victorian age their encounters with electricity – like those of the four bodies followed here – were distinctly corporeal. If we want to understand Victorian electricity, what it meant to the Victorians themselves and, quite literally sometimes, how they felt about it, we need to look at it through the lens of the Victorians' own bodies. The result, unsurprisingly, is a very different history of nineteenth-century electricity. But what this intimate link between Victorian bodies and electricity also suggests, I think, is that we cannot really understand how Victorians thought about their own bodies, their lives and deaths, without the electrical angle either. Electricity was many things to the Victorians: it was a dangerous and radical peril, it was a universal panacea, it was the power that governed bodies and minds, it was the force that held the empire together and it was the future. That it represented all those things (and a few others too) tells us a great deal about electricity in the nineteenth century – but tells us a great deal about the Victorians too.

Electricity at the beginning of the nineteenth century was hazardous material. It was indelibly associated with materialism and revolutionary fervour. It was the stuff of the mob, of demagoguery and quackery. But it could be made to work in other directions too. The experiments for which Tom Weems unwillingly consigned his body were an effort to wrest electricity from its radical roots – and this meant, perversely, trying to get away from its pervasive corporeality. That effort to get electricity away from the body continued throughout the Victorian period, but never quite succeeded, not even after the downfall of Cornelius Bennett Harness' electrical empire in the 1890s. That gentlemen of science such as Faraday took a jaundiced view of efforts to use electricity as a tool to understand body and soul, clearly did nothing to deter Ada Lovelace in her efforts at electrical self-exploration at the beginning of Victoria's reign. She shared the view of many radicals and popular electricians that electricity's bodily immediacy made it the ideal technology for experimenting with the apparatus of life. For Ada, as for others who shared her views, electricity was vitality. It explained the body and also subsumed the body in a broader culture of sensation and showmanship.

The spectacle and sensation that accompanied public electrical experimentation only served to emphasise electricity's apparent corporeality. Electricity was the science of shocks and sparks: things that were immediate and tangible, to the eye, the ear or the nervous system. We know that Constance Phipps' experience of electricity was intimately corporeal – it was practised on her body – but her experiences were also embedded in a broader culture of electrical performance. Electricity by the middle of Victoria's reign looked like the coming thing. It was the future on show and that electrical future was corporeal. It was about and for the body. The Medical Battery Company was in the business of selling the electrical future as well, with its promises of health through judicious doses of current. They were promising a new regime for managing the body. Their example reminds us of just how much late Victorian consumerism was about bodily matters. The bodies of middle-class *fin de siècle* Victorians were turning into battlegrounds for competing claims about the electrical future. Electricity in the century's last decade was still about sensation, but different visions of electricity incorporated competing views of how sensation should be managed. In the end it was all down to discipline, in all senses: how and where and by whom it should be imposed.

Victorian electricians playing with the body offered their audiences a vision of a future that was simultaneously liberating and threatening. They promised bodies that were free from all ills – but only if properly disciplined. Electricity would be an instrument for transforming bodies, nature and society; it might, according to some pundits, even be the means for defeating death. Electricity offered tools for dealing with individual and national degeneration as well as providing the raw matter for utopian dreaming. Electrical ideas and machinery were at the heart of Victorian aspirations to engineer themselves into the coming century. Their own bodies were the ultimate focus for those engineering efforts. In the end then, the human body itself was one of the most important sites for electrical experimentation for the Victorians. It was the primary medium through which electricity was assimilated into Victorian culture. Taking this corporeal element seriously means taking quite a different perspective on the places of physics throughout the nineteenth century. Electricity – and physics – was not just there in the laboratory or in the classroom, it leaked out wherever bodies were to be found as well. It titillated, it punished. Electricity helped the Victorians make sense of themselves and their place in the modern world they were in the process of creating around them.

NOTES

Part I Tom Weems

1 Tom Weems' Body

1 Roy Porter, *Enlightenment* (Harmondsworth: Penguin, 2001); Patricia Fara, *An Entertainment for Angels* (London: Icon Books, 2002).
2 Iain McCalman, *Radical Underworld* (Cambridge: Cambridge University Press, 1988).
3 John Rule, *Albion's People* (London: Longman, 1992); Pamela Horn, *The Rural World* (London: Hutchinson, 1980).
4 Martin Daunton, *Poverty and Progress* (Oxford: Oxford University Press, 1995).
5 Charles Johnson, *An Account of the Trials of the Ely and Littleport Rioters, in 1816* (Ely: C. Johnston, 1893).
6 Michael Murphy, *Cambridge Newspapers and Opinion, 1780–1850* (Cambridge: Oleander Press, 1977).
7 Peter Searby, *A History of the University of Cambridge*, vol. 2, (Cambridge: Cambridge University Press, 1997).
8 'A Shocking Murder', *The Times*, 17 May 1819, p. 3.
9 The inscription can be found on Mary Ann Weems' tombstone in the graveyard of St Mary's church, Godmanchester.
10 Peter King, *Crime, Justice and Discretion in England, 1740–1820* (Oxford: Oxford University Press, 2000); Douglas Hay, *Albion's Fatal Tree* (Harmondsworth: Penguin, 1988).
11 For Judge Burrough see his biography in the *Oxford Dictionary of National Biography* at: www.oxforddnb.com.
12 James Burrough to Lord Palmerston, 10 March 1822, in Harry Hopkins, *The Long Affray* (London: Secker & Warburg, 1985).
13 'Cambridge Assizes', *The Times*, 9 August 1819, p. 2.
14 Ibid.
15 'Trial for Murder', *Cambridge Chronicle*, 6 August 1819.
16 Ruth Richardson, *Death, Dissection and the Destitute* (London: Routledge & Kegan Paul, 1987).
17 'Execution of Weems', *Cambridge Chronicle*, 13 August 1819.
18 Ibid.
19 Ibid.
20 Ibid.
21 For Cumming see his biography in the *Oxford Dictionary of National Biography* at: www.oxforddnb.com; and William Brock, 'Coming and Going: the Fitful Career of James Cumming', in Mary Archer and Christopher Haley (eds), *The 1702 Chair of Chemistry at Cambridge* (Cambridge: Cambridge University Press, 2005).
22 Thomas Verney Okes, *An Account of the Providential Preservation of Elizabeth Woodstock, who Survived a Confinement Under the Snow of nearly Eight Days and Nights* (Cambridge: 1799).

23 Review of Thomas Verney Okes, 'An Account of Spina Bifida, with Remarks of a Method of Treatment proposed by Mr. Abernethy', in the *London Medical and Physical Journal*, 1810.

24 Arthur Rook, Margaret Carlton & W. Graham Cannon, *The History of Addenbrooke's Hospital, Cambridge* (Cambridge: Cambridge University Press, 1992).

25 Jan Golinski, *Science as Public Culture* (Cambridge: Cambridge University Press, 1992).

2 Galvanising Britain

1   Isaac Newton, *Opticks*, 4th edition (London: 1730), p. 328.

2   Simon Schaffer, 'Enlightened Automata', in William Clark, Jan Golinski & Simon Schaffer (eds), *The Sciences in Enlightened Europe* (Chicago IL: University of Chicago Press, 1999).

3   John Heilbron, *Electricity in the 17th and 18th Centuries* (Berkeley and Los Angeles CA: University of California Press, 1979).

4   Humphry Ditton, *The New Law of Fluids* (London: 1714), p. 41.

5   John Wesley, *The Desideratum, or Electricity made Plain and Useful* (London: 1760), p. 22.

6   Paola Bertucci, 'Revealing Sparks: John Wesley and the Religious Utility of Electrical Healing', *British Journal for the History of Science*, 2006, 39: pp. 341–62.

7   *Gentleman's Magazine*, 1745, 15: pp. 193–7.

8   Joseph Priestley, *History and Present State of Electricity* (London: 1767), p. 519.

9   The Venus Electrificata (or Venus Kiss) is described in Heilbron, *Electricity*, p. 267.

10 Heilbron, *Electricity*, p. 282.

11 Some of the ramifications are discussed in Paola Bertucci, 'The Electrical Body of Knowledge: Medical Electricity and Experimental Philosophy in the mid-18th Century', in Paola Bertucci & Giuliano Pancaldi (eds), *Electric Bodies: Episodes in the History of Medical Electricity* (Bologna: Universita di Bologna, 2001).

12 A recent and authoritative overview of Galvani, his context and impact is Marco Bresadola & Guilano Pancaldi (eds), *Luigi Galvani International Workshop Proceedings* (Bologna: Universita di Bologna, 1999).

13 For a detailed dissection of the debate between Galvani and Volta over animal electricity, see Marcello Pera, *The Ambiguous Frog* (Princeton NJ: Princeton University Press, 1992).

14 'Report Presented to the Class of the Exact Sciences of the Academy of Turin, 15th August 1802, in Regard to the Galvanic Experiments made by C. Vassali-Eandi, Giulio, and Rossi, on the 10th and 14th of the Same Month, on the Head and Trunk of Three Men a Short Time after their Decapitation', *Tilloch's Philosophical Magazine*, 1803, 15: pp. 38–45.

15 Volta's Parisian visit is discussed in Geoffrey Sutton, 'The Politics of Science in early Napoleonic France: the Case of the Voltaic Pile', *Historical Studies in the Physical Sciences*, 1981, 11: pp. 329–66.

16 'Galvanism', *Tilloch's Philosophical Magazine*, 1802, 14: pp. 191–2.

17 Ibid., pp. 364–6. The dashing Astley Cooper is also the subject of Druin Burch, *Digging up the Dead* (London: Chatto & Windus, 2007).

18 'Abstract of the Late Experiments of Professor Aldini on Galvanism', *Nicholson's Journal of Natural Philosophy*, 1802, 3: pp. 298–300.

19 Davy's enthusiastic adoption of the pile is described in Jan Golinski, *Science as Public Culture* (Cambridge: Cambridge University Press, 1992).

20 *The Times*, 22 January 1803, p. 3.

21 'George Forster', *Newgate Calendar*.

22 'Description of Mr. Pepys's Large Galvanic Apparatus', *Tilloch's Philosophical Magazine*, 1803, 15: pp. 94–6.

23 For the range of London entertainments during the first half of the nineteenth century see Richard Altick, *The Shows of London* (Cambridge MA: Belknap Press, 1978).

24 John [sic] Aldini, *General Views on the Application of Galvanism to Medical Purposes; Principally in Cases of Suspended Animation* (London: 1819), p. 26.

25 Ruth Richardson, *Death, Dissection and the Destitute* (London: Routledge & Kegan Paul, 1987).

26 Although it would not be long before Aldini was himself similarly lionised. See *Literary Journal*, 1804, 3: p. 383, which reported that 'Professor Aldini, so well known for his galvanic experiments, has received from the First Consul a large golden medal, on the one side of which is a head of Bonaparte, and on the other a laurel crown with an inscription tending to encourage him in the prosecution of his scientific researches. It was accompanied by a very flattering letter from the minister Chaptal.'

27 Joseph Priestley, *Experiments and Observations on Different Kinds of Air* (Birmingham: Thomas Pearson, 1790), p. xxiii.

28 Ann Thomson (ed.), *Machine Man and Other Writings* (Cambridge: Cambridge University Press, 1996), p. 40.

29 Ibid., p. 140.

30 Quoted in Walter Wetzels, 'Aspects of Natural Science in German Romanticism', *Studies in Romanticism*, 1971, 10: pp. 44–59, on p. 50.

31 Quoted in Walter Wetzels, 'Johann Wilhelm Ritter: Romantic Physics in Germany', Andrew Cunningham & Nick Jardine (eds), *Romanticism and the Sciences* (Cambridge: CUP, 1990), p. 203.

3 Galvanic Fashions

1 Charles Henry Wilkinson, *Elements of Galvanism, in Theory and Practice* (London: 1804), p. 319.

2 Described in John Bostock, *An Account of the History and Present State of Galvanism* (London: 1818), pp. 52–3.

3 Wilkinson, *Elements of Galvanism*.

4 Patricia Fara, *An Entertainment for Angels* (London: Icon Books, 2002).

5 'Description of Mr Pepys' large Galvanic Apparatus', *Tilloch's Philosophical Magazine*, 1803, 15: pp. 94–6.

6 For the Royal Institution's origins see Morris Berman, *Social Change and Scientific Organisation* (London: Heinemann, 1978).

7 Quoted in Henry Bence Jones, *The Royal Institution: Its Founder and First Professors* (London: 1871), p. 194.

8 Quoted in George Foote, 'Sir Humphry Davy and his Audience at the Royal Institution', *Isis*, 1952, 43: pp. 6–12, on p. 12.

9 Mike Jay, *The Atmosphere of Heaven: The Unnatural Experiments of Dr Beddoes and his Sons of Genius* (London: Yale University Press, 2009).

10 Thomas Beddoes (ed.), *Contributions to Physical and Medical Knowledge* (Bristol: Biggs & Cottle, 1799), p. 141.

11 John Davy, *Memoirs of the Life of Sir Humphry Davy* (London: 1836), vol. 1, p. 80.

12 Humphry Davy [writing anonymously], 'An Account of the Late Improvements in Galvanism ... By John Aldini', *Edinburgh Review*, 1803, 3: pp. 194–8.

13 Jan Golinski, *Science as Public Culture* (Cambridge: Cambridge University Press, 1992).

14 Charles Henry Wilkinson, *An Essay on the Leyden Phial, with a View to Explaining this Remarkable Phenomenon in Pure Mechanical Principles* (London: 1798), p. 60.

15 Roy Porter, *Flesh in the Age of Reason* (London: Allen Lane, 2003).

16 Wilkinson, *Elements of Galvanism*, p. 298.

17 Matthew Yatman, *Galvanism, Proved to be a Regular Assistant Branch of Medicine* (London: 1810), p. 3.

18 Wilkinson, *Elements of Galvanism*, p. 470.

19 Richard Holmes, *Coleridge: Early Visions* (London: Viking, 1990).

20 Ibid., p. 182.

21 Trevor Levere, *Poetry Realised in Nature* (Cambridge: Cambridge University Press, 1981).

22 Robert Southey to Davy, 26 July 1800, quoted in John Davy, *Fragmentary Remains, Literary and Scientific, of Sir Humphry Davy, Bart.* (London: 1858), p. 44.

23 Samuel Taylor Coleridge to Davy, 4 May 1801, quoted in Davy, *Fragmentary Remains*, p. 89.

24 Roy Porter, *Health for Sale* (Manchester: Manchester University Press, 1989).

25 Joseph Carpue, *An Introduction to Electricity and Galvanism* (London: 1803), p. 92.

26 Wilkinson, *Essay on the Leyden Phial*, p. 81.

27 James Delbourgo, *A Most Amazing Scene of Wonders: Electricity and Enlightenment in Early America* (Cambridge MA: Harvard University Press, 2006).

28 G.E. Mingay, *Mrs Hurst Dancing, and Other Scenes from Regency Life* (London. Gollancz, 1981).

29 Delbourgo, *A Most Amazing Scene of Wonders*.

30 Christopher Caustick, *The Modern Philosopher, or, Terrible Tractoration!* 2nd American edition, (Philadelphia: 1806), p. 130.

31 John Corry, *A Satirical View of London at the Commencement of the Nineteenth Century: Comprising Free Strictures on the Manners and Amusements of the Inhabitants of the English Metropolis; Observations on Literature and the Fine Arts; And Amusing Anecdotes of Public Characters*, 6th edn (London,: J. Ferguson, 1815), p. 60. (First edn, published 1801).

32 For some further insights into the Shelleys' knowledge of things galvanic see Sharon Ruston, *Shelley and Vitality* (London: Palgrave Macmillan, 2005).

33 Mary Shelley, *Frankenstein*, p. 30.

4 *Body & Soul*

1 The episode is described in Richard Holmes, *Coleridge: Early Visions* (London: Flamingo, 1999), pp. 135–68, quotations p. 159.

2 For Faraday and the City Philosophicals see Frank James, 'Michael Faraday, The City Philosophical Society and the Society of Arts', *Royal Society of Arts Journal*, 1992, 140: pp. 192–9.

3 John Thelwall, *The Rights of Nature Against the Usurpations of the Establishments* (Norwich: 1796), p. 4.

4 Edmund Burke, 'Letter to a Noble Lord', in *The Works of the Right Honourable Edmund Burke* (London: Holdsworth & Ball, 1834), vol. 2, p. 270.

5 Edmund Burke, 'Letters on a Regicidal Peace', in *The Works of the Right Honourable Edmund Burke* (London: Holdsworth & Ball, 1834), vol. 2, p. 282.

6 Pietro Corsi, *The Age of Lamarck* (Los Angeles & Berkeley CA: University of California Press, 1988).

7 Adrian Desmond, *The Politics of Evolution* (Chicago IL: University of Chicago Press, 1989).

8 George Macilwain, *Memoirs of John Abernethy* (London: Hurst & Blackett, 1853).

9 For the state and division of London medicine during the first half of the nineteenth century see M. Jeanne Peterson, *The Medical Profession in mid-Victorian London* (Berkeley & Los Angeles CA: University of California Press, 1978).

10 Peterson, *Medical Profession*, p. 26.

11 John Abernethy, *An Enquiry into the Probability and Rationality of Mr. Hunter's Theory of Life* (London: Longman, Hurst, Ress, Orme & Brown, 1814), pp. 16–7.

12 Abernethy, *An Enquiry*, p. 39.

13 See Roy Porter, *Flesh in the Age of Reason: The Modern Foundations of Body and Soul* (London: W.W. Norton & Co., 2003).

14 Michael Hawkins, 'A Great and Difficult Thing: Understanding and Explaining the
    Human Machine in Restoration England', in Iwan Rhys Morus (ed.), *Bodies/Machines*
    (Oxford: Berg Publications, 2002), pp. 15–38.

15 Albrecht von Haller, 'A Dissertation on the Sensible and Irritable Parts of Animals' (with
    an introduction by Owsei Temkin), *Bulletin of the History of Medicine and Allied Sciences*,
    1936, 4: pp. 651–99, on p. 658–9.

16 Abernethy, *An Enquiry*, p. 42.

17 Ibid., pp. 48–52.

18 See his biography in the *Oxford Dictionary of National Biography*, at www.oxforddnb.com.

19 William Lawrence, *An Introduction to Comparative Anatomy and Physiology* (London:
    J. Callow, 1816), pp. 120–1.

20 Ibid., p. 169.

21 Ibid., pp. 170–1. As we will see later, it turns out this was more or less exactly what one,
    at least, of Abernethy's supporters did want to say.

22 Ibid., p. 177.

23 John Abernethy, *Physiological Lectures, Exhibiting a General View of Mr. Hunter's Physiology*
    (London: Longman, Hurst, Rees, Orme & Brown, 1817), pp. 36–7.

24 William Lawrence, *Lectures on Physiology, Zoology, and the Natural History of Man* (London:
    J. Callow, 1819), p. 1.

25 Ibid., p. 6.

26 Ibid., p. 76.

27 John Abernethy, *The Hunterian Oration, for the Year 1819* (London: Strahan & Spottiswoode,
    1819), p. 59.

28 *The Radical Triumvirate, or, Infidel Paine, Lord Byron, and Surgeon Lawrence, Colleaguing with
    the Patriotic Radicals to Emancipate Mankind from all Laws Human and Divine* (London:
    Francis Westley, 1820).

29 Thomas Rennell, *Remarks on Scepticism, Especially as it is Connected with the Subjects of
    Organisation and Life, Being an Answer to the Views of M. Bichat, Sir T.C. Morgan, and Mr.
    Lawence* (London: F.C. & J. Rivington, 1819), pp. 1–2.

30 Rennell, *Remarks*, p. 8.

31 Ibid., p. 54.

32 Ibid., p. 64.

33 *Cursory Observations upon the Lectures on Physiology, Zoology, and the Natural History of Man,
    Delivered in the Royal College of Surgeons by W. Lawrence, in a Series of Letters Addressed to
    that Gentleman, with a Concluding Letter to his Pupils, by One of the People Called Christians*
    (London: T. Cadell & W. Davies, 1819), pp. 5–9.

34 Wylke Edwinsford, *A Review of a Work, Entitled Remarks on Scepticism, by the Rev. T.
    Rennell* (London: R. Carlile, 1819), pp. iii–iv.

35 Edwinsford, *Review*, p. 84.

36 'Abernethy, Lawrence &c. on the Theories of Life', *Quarterly Review*, 1819, 22: pp. 1–24,
    on pp. 4, 10.

37 'Modern Scepticism, as Connected with Organisation and Life', *British Review and
    London Critical Journal*, 1819, 14: pp. 169–205, on p. 170.

38 The title page described Edwinsford as a 'gentleman of Carmarthenshire'. I have
    been unable to trace him further, though there certainly was a contemporary gentry
    Carmarthernshire family of that name. Until recently, the little village of Talyllychau
    posessed a pub called the Edwinsford Arms.

39 'Recent Controversy on Materialism', *Monthly Repository of Theology and General
    Literature*, 182, 17: pp. 170–82, on p.170.

40 Thomas Ashe (ed.), *Miscellanies, Aesthetic and Literary, to which is Added the Theory of Life, by
    Samuel Taylor Coleridge* (London: George Bell & Sons, 1892), p. 405.

41 Ibid., p. 375.

42 Ibid., pp. 378–86.

5 Dissecting Tom Weems

1 Andrew Ure, 'An Account of Some Experiments made on the Body of a Criminal immediately after Execution, with Physiological and Practical Observations', *Quarterly Journal of Science*, 1819, 6: pp. 283–94.

2 W.V. Farrar, 'Andrew Ure FRS and the Philosophy of Manufactures', *Notes & Records of the Royal Society*, 1973, 27: pp. 299–324.

3 Peter Mackenzie, *Old Reminiscences of Glasgow and the West of Scotland* (Glasgow: James P. Forrester, 1890), vol. 2, p. 492.

4 Ibid., p. 496.

5 'Execution at Glasgow', *The Times*, 11 November 1818, p. 3.

6 Ure, 'An Account', pp. 289–90.

7 Ibid., pp. 290–3.

8 Ibid., p. 290.

9 'Horrible Phenomena – Galvanism!', *The Times*, 11 February 1819, p. 3.

10 Mackenzie, *Old Reminiscences*, p. 499.

11 'Execution of Weems', *Cambridge Chronicle*, Friday 13 August 1819.

12 Ibid.

13 'Execution', *The Times*, 27 July 1824, p. 3.

14 'Spirit of Discovery', *Mirror of Literature, Amusement and Instruction*, 1830, 15: pp. 95–6.

15 Andrew Ure, *The Philosophy of Manufactures* (London: 1835), pp. 9–13.

16 Karl Marx, *Capital*, vol. 1 (Harmondsworth: Penguin, 1982).

Part II Ada Lovelace

6 Knowing Ada

1 Undated fragment, in Betty Alexandra Toole (ed.), *Ada, The Enchantress of Numbers* (Mill Valley CA: Strawberry Press, 1992), p. 293.

2 Ada Lovelace to Woronzo Grieg, in ibid., p. 297.

3 Alison Winter, 'A Calculus of Suffering: Ada Lovelace and the Bodily Constraints on Women's Knowledge in early Victorian England', in Christopher Lawrence & Steven Shapin (eds), *Science Incarnate* (Chicago: University of Chicago Press, 1998), pp. 202–39.

4 Cornelia Crosse, *Red-letter Days of my Life*, (London: Richard Bentley & Son, 1892), vol. 1, pp. 22–3.

5 Cornelia Crosse, *Memorials, Scientific and Literary, of Andrew Crosse, the Electrician* (London: Longman, Brown, Green, Longman & Roberts, 1857), p. 9.

6 Dorothy Stein, *Ada: A Life and a Legacy* (Cambridge MA: MIT Press, 1985), p. 14.

7 Henry B. Stanton, *Sketches of Reforms and Reformers of Great Britain and Ireland* (New York NY: John Wiley, 1849), p. 353.

8 Stein, *Ada*, p. 18.

9 Ibid., p. 22.

10 Ibid., p. 24.

11 Ibid., p. 25.

12 Ibid., p. 36.

13 Ibid., p. 34.

14 Ada Byron to Mary Somerville, 20 February 1835, in Toole (ed.), *Ada, The Enchantress*, p. 72.

15 Ada Lovelace to Annabella Byron, undated, in ibid., p. 101.

16 Ada Byron to William King, 9 March 1834, in ibid., p. 53.

17 Quoted in Stein, *Ada*, p. 43. The study of mathematics was widely regarded as a tool for disciplining the mind. See also Andrew Warwick, *Masters of Theory* (Chicago: University of Chicago Press, 2003).

18 Ada Byron to William King, 9 March 1834, in Toole (ed.), *Ada, The Enchantress*, p. 53.

19 Crosse, *Red-letter Days*, vol. 1, p. 169.

20 AAL [Ada Lovelace], 'Sketch of the Analytical Engine by L.F. Menabrea', *Taylor's Scientific Memoirs*, 1843, 3: pp. 666–731.

21 Ada Lovelace to Charles Babbage, 5 July 1843, in Toole (ed.), *Ada, The Enchantress*, p. 203.

22 Cornelia Crosse, 'Science and Society in the Fifties', *Temple Bar*, 1891, 93: pp. 33–51, on p. 48.

23 Crosse, *Red-letter Days*, vol. 1, p. 15.

24 Ibid., p. 25.

25 Crosse, *Memorials*, p. 59.

26 Ibid., p. 42.

27 Crosse, *Red-letter Days*, p. 24.

28 Crosse, *Memorials*, p. 61.

29 Jack Morrell & Arnold Thackray, *Gentlemen of Science: The Early Years of the British Association for the Advancement of Science* (Oxford: Oxford University Press, 1981).

30 Sidney Ross, 'Scientist: The Story of a Word', *Annals of Science*, 1962, 18: pp. 65–85.

31 Richard Phillips, 'A Brief Account of a Visit to Andrew Crosse, Esq., of Broomfield', *Annals of Electricity*, 1836–37, 1: pp. 135–45, on p. 135.

32 Crosse, *Memorials*, pp. 146–50.

33 Phillips, 'Brief Account', p. 142.

34 Ibid., p. 138.

35 Ibid., pp. 140–2.

36 Andrew Crosse, 'Description of Some Experiments Made with the Voltaic Battery, by Andrew Crosse, Esq., of Broomfield, near Taunton, for the Purpose of Producing Crystals; in the Process of Which Experiments Certain Insects Constantly Appeared', *Annals of Electricity*, 1838, 2: pp. 246–57.

37 Crosse, *Memorials*, p. 354.

38 James A. Secord, 'Extraordinary Experiment: Electricity and the Creation of Life in Victorian England', in David Gooding, Trevor Pinch & Simon Schaffer (eds), *The Uses of Experiment* (Cambridge: Cambridge University Press, 1989), pp. 337–83. See also Oliver Stallybrass, 'How Faraday Produced Living Animalculae: Andrew Crosse and the Story of a Myth', *Proceedings of the Royal Institution*, 1967, 41: pp. 597–619.

39 Stephen Wheeler, *The Poetical Works of Walter Savage Landor* (Oxford: Clarendon Press, 1937), vol. 1, p. 431.

40 Ada Lovelace to William Lovelace, 22 November 1844, in Toole (ed.), *Ada, The Enchantress*, p. 297.

41 Ada Lovelace to William Lovelace, 25 November 1844, in ibid., p. 300.

42 Ada Lovelace to William Lovelace, 29 November 1844, in ibid., p. 302.

43 Ada Lovelace to William Lovelace, 24 November 1844, in ibid., p. 298.

44 Ada Lovelace to William Lovelace, 29 November 1844, in ibid., p. 301.

45 Ada Lovelace to Woronzow Greig, 15 November 1844, in ibid., p. 295.

46 Ada Lovelace to Andrew Crosse, undated, in ibid., p. 296.

7 *Electric Universe*

1 On Faraday see L. Pearce Williams, *Michael Faraday* (London: Chapman & Hall, 1965) and Iwan Rhys Morus, *Michael Faraday and the Electrical Century* (London: Icon Books, 2004).

2 Geoffrey Hubbard, *Cooke, Wheatstone and the Invention of the Electric Telegraph* (London: Routledge & Kegan Paul, 1965).

3 *Patent Journal and Inventor's Advocate*, 1850, vol. 10, p. iv.

4 Alfred Smee, *Elements of Electrometallurgy*, 3rd edition, (London: 1844), p. 348.

5 William Fothergill Cooke, *The Electric Telegraph: Was it Invented by Professor Wheatstone?* (London: 1856–57), vol. 2, p. 251. For electrical enthusiasms in general see Iwan Rhys Morus, *Frankenstein's Children: Electricity, Exhibition and Experiment in early Nineteenth-century London* (Princeton NJ: Princeton University Press, 1998).

6 George Eliot, *The Mill on the Floss*, first published 1860. Great Writers edition, (London: Marshall Cavendish, 1986), p. 289.

7 Lady Jane Pollock [writing anonymously], 'Michael Faraday', *St Paul's Magazine*, 1870, 4, pp. 293–303.

8 Henry Bence Jones, *Life and Letters of Faraday* (London: 1870), vol. 2, p. 441.

9 Quoted in Morus, *Frankenstein's Children*, p. 77.

10 'The Royal Polytechnic Institution', *Morning Chronicle*, 16 September 1843.

11 Morus, *Frankenstein's Children*, chapter 4.

12 Henry M. Noad, *A Course of Eight Lectures on Electricity, Galvanism, Magnetism, and Electro-magnetism* (London: 1844), Dedication.

13 Auguste de la Rive, 'Some Notes on the Present State of the Study of Electricity in England, Collected During a Recent Sojourn in that Country', *Electrical Magazine*, 1845, 1: pp. 100–7.

14 Charles V. Walker, 'Mr. Gassiot's Electrical Soirée', *Electrical Magazine*, 1845, 1: p. 537.

15 William Whewell, *The Philosophy of the Inductive Sciences*, (London: 1840), vol. 1, p. 335.

16 William Robert Grove, *On the Correlation of Physical Forces* (London: London Institution, 1846), pp. 7–8.

17 Michael Faraday, 'Fifteenth Series', *Experimental Researches in Electricity* (London: Taylor & Francis, 1839–52), vol. 2, pp. 1–127, on p. 1.

18 Ibid., p. 15.

19 William Sturgeon, 'On Electro-magnetism', *Philosophical Magazine*, 1824, 64: pp. 242–9, on p. 248.

20 William Sturgeon, 'A General Outline of the Various Theories which have been Advanced for the Explanation of Terrestrial Magnetism', *Annals of Electricity*, 1836–37, 1: pp 117–23, on p. 123.

21 William H. Weekes, *Proceedings of the London Electrical Society*, 1841.

22 Robert Were Fox, 'Theoretical Views of the Origins of Mineral Veins', *Annals of Electricity*, 1838, 2: pp. 166–94, on p. 175.

23 Thomas Pine, 'On the Probable Connection between Electricity and Vegetation', *Mechanics' Magazine*, 1835–36, 24: pp. 99–104, on p. 101.

24 Francis Maceroni, *Memoirs and Adventures of Colonel Maceroni, Late Aide-de-Camp to Joachim Murat, King of Naples*, 2 vols (London: 1838).

25 Francis Maceroni, 'An Account of Some Remarkable Phenomena seen in the Mediterranean, with Some Physiological Deductions', *Mechanics' Magazine*, 1831, 15: pp. 93–6.

26 Eliza Sharples, 'An Inquiry How Far the Human Character is Formed by Education or External Circumstances', *The Isis*, 1832, 1: pp. 81–5, on p. 85.

27 Eliza Sharples, 'Fifth Discourse on the Bible', *The Isis*, 1832, 1: pp. 241–7.

28 Thomas Simmons Mackintosh, *The Electrical Theory of the Universe* (Boston: 1846), p. 90.

29 Ibid., p. 360.

30 Ibid., p. 371.

31 Patrick Murphy, *Rudiments of the Primary Forces of Gravity, Magnetism, and Electricity, in their Agency on the Heavenly Bodies* (London: 1830), p. 58.

32 William Leithead, *Electricity; Its Nature, Operation and Importance in the Phenomena of the Universe* (London: Longman, Orme, Brown, Green & Longman, 1837), p. 1.
33 Ibid., p. 394.
34 Ibid., p. 395.
35 James Secord (ed.), *Vestiges of the Natural History of Creation and Other Evolutionary Writings* by Robert Chambers (Chicago IL: University of Chicago Press, 1994), p. 163.
36 Ibid., p. 334.
37 James Secord, *Victorian Sensation: The Extraordinary Publication, Reception and Secret Authorship of Vestiges of the Natural History of Creation* (Chicago: University of Chicago Press, 2000).
38 Alfred Smee, *Elements of Electrobiology* (London: 1848).
39 Alfred Smee, *Instinct and Reason, Deduced from Electro-biology* (London: 1850), p. 274.
40 'Instinct and Reason', *British & Foreign Medico-chirurgical Review*, 1850, 6: pp. 522–4.

8 *Galvanic Medicine*

1   William Sturgeon, *A Course of Twelve Elementary Lectures on Galvanism* (London: Sherwood, Gilbert & Piper, 1843), p. v.
2   Marshall Hall, 'On the Functions of the Medulla Oblongata and Medulla Spinalis, and on the Excito-motory System of Nerves', *Proceedings of the Royal Society*, 1830–37, 3: pp. 463–4.
3   For the coup at the Royal Society see Marie Boas Hall, *All Scientists Now: The Royal Society in the Nineteenth Century* (Cambridge: Cambridge University Press, 1984) and Iwan Rhys Morus, 'Correlation and Control: William Robert Grove and the Construction of a New Philosophy of Scientific Reform', *Studies in History & Philosophy of Science*, 1991, 22: pp. 589–621.
4   S. Wilks & G.T. Bettany, *A Biographical Dictionary of Guy's Hospital* (London: 1892), p. 429.
5   Thomas Williams, 'On the Laws of the Nervous Force, and the Function of the Roots of the Spinal Nerve', *Lancet*, 1847, II: pp. 516–7.
6   Charles Vincent Walker, 'Notices of New Books', *Electrical Magazine*, 1845, 1: pp. 542–50.
7   Golding Bird, 'Report on the Value of Electricity as a Remedial Agent in the Treatment of Disease', *Guy's Hospital Reports*, 1841, 6: pp. 84–120, on p. 84.
8   Ibid., p. 85.
9   Ibid., p. 87.
10 Ibid., pp. 88–9.
11 H.M. Hughes, 'Digest of One Hundred Cases of Chorea, Treated in the Hospital', *Guy's Hospital Reports*, 1846, 4: pp. 360–94; William W. Gull, 'A Further Report on the Value of Electricity as a Remedial Agent', *Guy's Hospital Reports*, 1852, 8: pp. 81–143.
12 Ibid., p. 82.
13 Golding Bird, 'On the Employment of Electro-magnetic Current in the Treatment of Paralysis', *Lancet*, 1846, I: pp. 649–51.
14 Ibid., p. 651.
15 William H. Halse, 'Wonderful Effects of Voltaic Electricity in Restoring Animal Life when the Sensorial Powers have Entirely Ceased or in other Words, when Death in the Common Acceptation of the Term has Actually Occurred', *Annals of Electricity*, 1840, 4: pp. 481–4.
16 'Grand Display of Animal Magnetism!' Broadsheet held at Bakken Library and Museum of Electricity in Life, Minneapolis (dated 1843).
17 William Hooper Halse, *The Extraordinary Remedial Efficacy of Medical Galvanism, when Scientifically Administered* (London: n.d.), p. 1.
18 Ibid., p. 31.
19 Dawson Bellhouse, *Ten Minutes Reading on Medical Galvanism and its Properties* (Liverpool: n.d.), p. 16.

20 Halse, *Medical Galvanism*, p. 17.

21 Charles V. Walker, 'Facts (?) in the History of Electricity', *Electrical Magazine*, 1845, 1: pp. 551–2.

22 Doggrel Drydog, 'Gals and Galvanism!', Pamphlet broadsheet.

23 W.P. Piggot, *Galvanic Belts, and Galvanism* (London: 1852), p. 2.

24 J.O.N. Rutter, *Human Electricity: The Means of its Development, Illustrated by Experiments* (London: 1854), p. 167.

25 Ibid., p. 125.

26 Ibid., p. 177.

27 William Leithead, *Electricity; Its Nature, Operation, and Importance in the Phenomena of the Universe* (London: Longman, Orme, Brown, Green & Longman, 1837), p. 332.

28 Ibid., p. 332.

29 Anon., 'Reviews and Notices Respecting New Books', *Philosophical Magazine*, 1838, 11: pp. 127–30.

30 Leithead, *Electricity*, p. 380.

31 James Murray, *Electricity as a Cause of Cholera, or Other Epidemics, and the Relation of Galvanism to the Action of Remedies* (Dublin: 1849).

32 J.G. Milligan, *Mind and Matter* (London: 1847), p. 157.

33 James Henry Bennet, *A Practical Treatise on Inflammation of the Uterus* (London: John Churchill, 1853), p. 337. First edn 1845.

34 Marshall Hall, *Lectures on the Nervous System and its Diseases* (Philadelphia: E.L. Carey & A. Hart, 1836), p. 219.

35 Marshall Hall, *Essays Chiefly on the Theory of the Paroxysmal Diseases of the Nervous System* (London: 1849), p. 23.

36 Edward John Tilt, *On Diseases of Women and Ovarian Inflammation* (London: 1853), p. 9.

37 Ibid., p. 73.

38 Ibid., p. 56.

39 Thomas William Nunn, *Inflammation of the Breast, and Milk Abscess* (London: 1853), p. 4.

40 Robert Brudenell Carter, *On the Pathology and Treatment of Hysteria* (London: 1853), p. 36.

41 Ibid., p. 67.

42 Thomas Laycock, *An Essay on Hysteria* (Philadelphia: 1840), p. 59.

43 Ibid., p. 72.

44 Ibid., p. 76.

45 Thomas Laycock, *A Treatise on the Nervous Diseases of Women* (London: 1840), p. 142.

46 Thomas Laycock, 'On the Treatment of Cerebral Hysteria, and of Moral Insanity in Women, by Electro-galvanism', *Medical Times*, 1850, 1: pp. 57–8.

47 Thomas Laycock [writing anonymously], 'Woman in her Psychological Relations', *Journal of Psychological Medicine & Mental Pathology*, 1851, 4: pp. 18–50, on p. 38.

48 Laycock, 'Cerebral Hysteria', p. 57.

49 Ibid., p. 58.

50 Golding Bird, *Lectures on Electricity and Galvanism, in their Physiological and Therapeutical Relations* (London: 1849), p. 163.

51 Ibid., p. 148.

9 *Ada's Laboratory*

1 Ada Lovelace to Woronzow Greig, 10 February 1844, in Betty Alexandra Toole (ed.), *Ada, the Enchantress of Numbers* (Mill Valley CA: Strawberry Press, 1992), p. 270.

2 Ada Lovelace to Lady Byron, 11 November 1844, in ibid., p. 291.

3 Ibid.

4 Ada Lovelace to Andrew Crosse, undated, in ibid., p. 296.

5 Ada Lovelace to Lady Byron, undated, in ibid., p. 288.

6   Boyd Hilton, *The Age of Atonement: The Influence of Evangelicalism on Social and Economic Thought, 1785–1865* (Oxford: Oxford University Press, 1992).

7   Ada Lovelace to Michael Faraday, 16 October 1844, in Frank James (ed.), *The Correspondence of Michael Faraday*, vol. 3 (London: IEE, 1996), p. 253.

8   Ibid., p. 254.

9   Michael Faraday to Ada Lovelace, 24 October 1844, in ibid., p. 265.

10 Ibid., p. 266.

11 Ada Lovelace to Michael Faraday, 24 October 1844, in ibid., p. 267.

12 Duke of Somerset to Charles Babbage, 16 December 1835, in Morris Berman, *Social Change and Scientific Organisation* (London: Heinemann, 1978), p. 174.

13 John Tyndall, *Faraday as a Discoverer* (London: 1868).

14 Michael Faraday, 'On Mental Education', *Experimental Researches in Chemistry and Physics* (London: Richard Taylor & William Francis, 1859), p. 464.

15 Michael Faraday, 'On Table-turning', *Experimental Researches in Physics and Chemistry*, p. 383.

16 Robert Darnton, *Mesmerism and the End of the Enlightenment in France* (Cambridge MA: Harvard University Press, 1967); Roger Cooter, *The Cultural Meaning of Popular Science: Phrenology and the Organization of Consent in Nineteenth-century Britain* (Cambridge: Cambridge University Press, 1984).

17 Alison Winter, *Mesmerized* (Chicago: University of Chicago Press, 1998).

18 Undated fragment in Toole (ed.), *Ada, The Enchantress*, p. 282; and Ada Lovelace to Lady Byron, 10 October 1844, p. 282.

19 Alison Winter, 'Harriet Martineau and the Reform of the Invalid', *Historical Journal*, 1995, 38: pp. 597–616.

20 Will Ashworth, 'The Calculating Eye: Baily, Herschel, Babbage and the Business of Astronomy', *British Journal for the History of Science*, 1994, 27: pp. 409–41.

21 Charles Babbage, *Reflexions on the Decline of Science in England* (London: 1830).

22 Charles Babbage, *Passages from the Life of a Philosopher* (London; Longman, Green, Longman, Roberts & Green, 1864), p. 31.

23 John Herschel, *Preliminary Discourse on the Study of Natural Philosophy* (London: Longman, Rees, Orme, Brown & Green, 1830), p. 70.

24 Charles Babbage, *The Ninth Bridgewater Treatise: A Fragment* (London: John Murray, 1837).

25 Ada Lovelace to Lord Lovelace, 24 November 1844, in Toole (ed.), *Ada, The Enchantress*, p. 299.

26 Ada Lovelace to Lord Lovelace, 29 November 1844, in ibid., p. 301.

Part III Constance Phipps

10 *Lady Constance's Pain*

1   Iwan Rhys Morus, *Frankenstein's Children* (Princeton NJ: Princeton University Press, 1998), chapter 8.

2   George Eland, *The Lobb Family from the Sixteenth Century* (Oxford: Oxford University Press, 1955), p. 92.

3   Ibid.

4   Roy Porter, *London: A Social History* (London: Hamish Hamilton, 1995); Lynda Nead, *Victorian Babylon* (London: Yale University Press, 2005).

5   Both letters are undated but were clearly written sometime in the later 1850s (but before 1859 since they refer to the earlier, Gloucester Terrace, address). They are part of a small collection of Lobb correspondence in private hands. I am deeply indebted to Andrew Lobb and Godfrey Cory-Wright for their assistance in tracing this material and allowing me to make use of it.

6   Harry Lobb, *On the Curative Treatment of Paralysis and Neuralgia, and Other Affections of the Nervous System with the Aid of Galvanism* (London: Hippolyte Bailllere, 1859), dedication.

7   A notice in the *Electrician*, 6 November 1863, gives a sample list of patrons.

8   Ibid.

9   'The Present State of the Science of Medical Electricity', *Electrician*, 1863, 5: p. 6.

10 'Harveian Society of London, February 19, 1863. Henry W. Fuller M.D., President, in the Chair. "On the Uses and Value of Galvanism and Electricity in General Practice." By Harry Lobb Esq.', *British Medical Journal*, 1863, I: pp. 226–7.

11 Royal College of Surgeons, Council Minutes, vol. XII, pp. 128–69. See Iwan Rhys Morus, 'Batteries, Bodies and Belts: Making Careers in Victorian Medical Electricity', Paola Bertucci & Giuliano Pancaldi (eds), *Electric Bodies: Episodes in the History of Medical Electricity* (Bologna: Universita di Bologna, 2001), pp. 209–38

12 'Medical Advertisements in the Daily Journals', *British Medical Journal*, 1874,I: p. 327; 'Harveian Society of London', *British Medical Journal*, 1874, I: p. 355; 'Harveian Society of London', *British Medical Journal*, 1874, I: p. 419.

13 Again, the letters are undated so it is impossible to date then definitively, or even be sure of the order in which they were written. They are marked Jan 27, June 9 and Oct 14 but whether these are all the same year it is difficult to tell.

14 Hector Colwell, *An Essay in the History of Electrotherapy and Diagnosis* (London: Heinemann, 1922).

15 Lobb, *Paralysis and Neuralgia*, p. 13.

16 *Practical Guide for the Electro-medical Treatment of Rheumatic and Nervous Diseases, Gout, etc., by Means of Pulvermacher's Hydro-electric Chains* (London: Pulvermacher's General Depot, 1856), p. 6.

17 Ibid., p. 1.

18 Ibid., p. 10.

19 Lobb, *Paralysis and Neuralgia*, pp. 76–7.

20 Harry Lobb, 'Acute Neuralgia Cured with the Aid of the Continuous Current, with Peculiar Sympathetic Effect upon the Uterus', *Lancet*, 1860, I: p. 526.

21 Herbert Tibbits, *How to Use a Galvanic Battery in Medicine and Surgery* (London: 1877), p. 1.

22 Julius Althaus, *Treatise on Medical Electricity* (London: 1859), p. 344.

23 Tibbits, *Galvanic Battery*, p. 46.

24 'The March of Specialism', *Lancet*, 1863, I: p. 183.

25 Harry Lobb, 'Special Hospitals – the London Galvanic Hospital', *Lancet*, 1863, I: p. 219.

11 Electric Frontier

1   Ken Beauchamp, *Exhibiting Electricity* (London: Institute of Electrical Engineers, 1997).

2   Edward Bulwer Lytton, *The Coming Race* (London: Alan Sutton Publishing, 1995), first edn 1871, p. 20.

3   Leslie Mitchell, *Bulwer Lytton: The Rise and Fall of a Victorian Man of Letters* (London: Hambledon & London, 2003).

4   Bulwer Lytton, *Coming Race*, p. 20.

5   Bruce Hunt, 'The Ohm is where the Art is: British Telegraph Engineers and the Development of Electrical Standards', *Osiris*, 1994, 9: pp. 48–63; Iwan Rhys Morus, 'The Electric Ariel: Telegraphy and Commercial Culture in early Victorian England', *Victorian Studies*, 1996, 29: pp. 339–78.

6   W.J. Copleston, *Memoir of Edward Copleston, D.D., Bishop of Llandaff* (London: 1851), p. 169.

7   Andrew Wynter, 'The Electric Telegraph', *Quarterly Review*, 1854, 95: p. 119.

8   Ibid., p. 132.

9   George Wilson, *Electricity and the Electric Telegraph* (London: 1855), p. 77.

10 G.B. Prescott, *History, Theory and Practice of the Electric Telegraph* (Boston: 1860), p. 242.

11 Graeme Gooday, *Domesticating Electricity: Technology, Uncertainty and Gender, 1880–1914* (London: Pickering & Chatto, 2008).

12 'Davenport's Electro-magnetic Engine', *Mechanic's Magazine*, 1837, 27: pp. 404–5.

13 Frederick C. Bakewell, *Electric Science; its History, Phenomena and Applications* (London: Ingram, Cooke & Co., 1853), p. 191.

14 Beauchamp, *Exhibiting Electricity*, p. 98.

15 'The International Exhibition', *The Times*, 3 June 1862, p. 14.

16 'The International Exhibition', *The Times*, 10 June 1862, p. 9.

17 'The International Exhibition', *The Times*, 17 June 1862, p. 11.

18 'The International Exhibition', *The Times*, 23 June 1862, p. 9.

19 'The Illuminations', *The Times*, 11 March 1863, p. 7.

20 Editorial, *The Times*, 11 March 1863, p. 9.

21 'New Electric Light', *The Times*, 3 August 1860, p. 12.

22 'Laying the Ghost', *The Times*, 10 April 1863, p. 12.

23 'The Opening of the Middle Temple Library', *The Times*, 1 November 1861, p. 7.

24 Bernard Lightman, *Victorian Popularizers of Science: Designing Nature for New Audiences* (Chicago: University of Chicago Press, 2007).

25 John Henry Pepper, *The Boy's Playbook of Science* (London: 1866).

26 John Henry Pepper, *The True History of the Ghost; and all about Metempsychosis* (London: Cassell & Co., 1890), p. 3.

27 Iwan Rhys Morus, 'More the Aspect of Magic than Anything Natural: The Philosophy of Demonstration', Bernard Lightman & Aileen Fyfe (eds), *Science in the Marketplace: Nineteenth-century Sites and Experiences* (Chicago: University of Chicago Press, 2007), pp. 336–70.

28 'The Great Induction Coil at the Polytechnic Institution', *The Times*, 7 April 1869, p. 4.

29 Ibid.

30 Benjamin Ward Richardson, *Hygeia: The City of Health* (London: 1876).

31 Benjamin Ward Richardson, 'On Research with the Large Induction Coil of the Royal Polytechnic Institution, with Special Reference to the Cause and Phenomena of Death by Lightning', *Medical Times and Gazette*, 1869, 38: pp. 39, 183-6, 511–4, 595–9; on p. 598. Tobin was one of Pepper's assistants at the Royal Polytechnic Institution.

32 John P. Gassiot, 'On Some Experiments made with Ruhmkorff's Induction Coil', *Philosophical Magazine*, 1854, 7: pp. 97–9.

33 Henry M. Noad, *The Inductorium, or Induction Coil* (London: 1868), p. 59.

34 Ibid., p. 58.

35 Ibid., pp. 68–9.

36 Robert Strutt, *The Life of Lord Rayleigh* (London: Arnold, 1924).

37 Quoted in Lewis Campbell & William Garnett, *The Life of James Clerk Maxwell* (London: 1882), p. 270.

38 James Clerk Maxwell [writing anonymously], 'Electricity and Magnetism', *Nature* 15 May 1873, pp. 42–3.

39 Fleeming Jenkin, *Electricity and Magnetism* (London: 1873), pp. v–vi.

40 Maxwell, 'Electricity and Magnetism', p. 42.

12 *Machinery of the Body*

1   Andrew King & John Plunkett, *Victorian Print Media* (Oxford: Oxford University Press, 2005).

2   Anthony Failing, *A History of the Maida Vale Hospital for Nervous Diseases* (London: Butterworth, 1958).

3   Julius Althaus, *A Treatise on Medical Electricity* (London: 1870), 2nd edn, p. 431.

4   Ibid., p. 432.

5   Ibid., p. 347.

6   Julius Althaus, 'The Functions of the Brain', *Nineteenth Century*, 1879, 6: pp. 101–32.

7   Guilleme-Benjamin Duchenne, *The Mechanism of Human Facial Expression*, edited and translated by R. Andrew Cuthbertson (Cambridge: Cambridge University Press, 1990).

8   Herbert Tibbits, *Handbook of Medical and Surgical Electricity* (London: J.A. Churchill, 1877), 2nd edn, p. 114.

9   Harry Lobb, *On Some of the More Obscure Forms of Nervous Affections: Their Pathology and Treatment* (London: 1858), p. 297.

10 Ibid., p. 238.

11 Harry Lobb, *A Popular Treatise on Curative Electricity, Especially Addressed to Sufferers from Paralysis, Rheumatism, Neuralgia, and Loss of Nervous & Physical Power* (London: 1867), p. 36.

12 Harry Lobb, *On Some of the More Obscure Forms of Nervous Affections: Their Pathology and Treatment*, 2nd edn (London: 1863), p. 193.

13 Janet Oppenheim, *Shattered Nerves: Doctors, Patients and Depression in Victorian England* (Oxford: Oxford University Press, 1991); Michael Mason, *The Making of Victorian Sexuality* (Oxford: Oxford University Press, 1994).

14 Boyd Hilton, *A Mad, Bad and Dangerous People: England 1783–1846* (Oxford: Oxford University Press, 2006).

15 William Robert Grove, *On the Correlation of Physical Forces* (London: 1846); Iwan Rhys Morus, 'Correlation and Control: William Robert Grove and the Construction of a New Philosophy of Scientific Reform', *Studies in History and Philosophy of Science*, 1991, 22: pp. 589–621.

16 William Benjamin Carpenter, 'On the Mutual Relations of the Vital and Physical Forces', *Philosophical Transactions*, 1850, 140: pp. 727–57.

17 Harry Lobb, *On the Curative Treatment of Paralysis and Neuralgia, and other Affections of the Nervous System with the Aid of Galvanism* (London: 1859), pp. 5–6.

18 Joseph Snape, *Electro-Dentistry: Facts, and Observations* (London: 1865), p. 22.

19 William Robert Grove, 'An Address on the Importance of Physical Science in Medical Education', *British Medical Journal*, 1869, 1: pp. 485–7.

20 Lobb, *Nervous Affections*, p. 73.

21 Harry Lobb, 'An Attempt to Adapt the Laws of Electro-dynamics to the Science of Medical Electricity', *British & Foreign Medico-chirurgical Review*, 1866, 73: pp. 508–16.

22 Peter Guthrie Tait to William Thomson, 2 December 1862, Thomson Papers, Glasgow University Library. Tait had just finished reading the 4th edition of *Correlation*, published in 1862.

23 James Clerk Maxwell to Michael Faraday, 9 November 1857, in L. Pearce Williams (ed.), *The Selected Correspondence of Michael Faraday* (Cambridge: Cambridge University Press, 1971), vol. 2, p. 881.

24 Balfour Stewart, *The Conservation of Energy: Being an Elementary Treatise on Energy and its Laws* (London: Henry S. King & Co., 1873), pp. 161–2.

25 Henry Maudsley, 'Sex in Mind and Education', *Fortnightly Review*, 1874, 15: pp. 479–80.

26 Oppenheim, *Shattered Nerves*; Cynthia Eagle Russet, *Sexual Science: The Victorian Construction of Womanhood* (Cambridge MA: Harvard University Press, 1981).

27 Crosbie Smith, *The Science of Energy: A Cultural History of Energy in Victorian Britain* (London: Athlone, 1998).

28 John Tyndall, *Address Delivered before the British Association Assembled at Belfast, with Additions* (London: Longman, Green & Co., 1874), p. 59.

29 Thomas Henry Huxley, 'On the Hypothesis that Animals are Automata, and its History', *Fortnightly Review*, 1874, 16: pp. 555–80.

30 Ibid., p. 577–8.
31 William Benjamin Carpenter [writing anonymously], 'On the Doctrine of Human Automatism', Metaphysical Society Papers, read Tuesday 17 November at the Grosvenor Hotal, at 8.30 p.m.
32 William Benjamin Carpenter, *Is Man an Automaton? A Lecture Delivered in the City Hall, Glasgow, in 23 February 1875* (London: 1865), pp. 5–6.
33 William Benjamin Carpenter, 'On the Doctrine of Human Automatism', *Contemporary Review*, 1875, 25: pp. 397–416, on p. 407.
34 William Benjamin Carpenter, 'On the Doctrine of Human Automatism, Part II', *Contemporary Review*, 1875, 25: pp. 940–62, on p. 945.
35 William Benjamin Carpenter, 'On the Physiological Import of Dr. Ferrier's Experimental Investigations into the Functions of the Brain', *West Riding Lunatic Asylum Medical Reports*, 1874, 4: pp. 1–23, on p. 3.
36 William Benjamin Carpenter, 'The Limits of Human Automatism', *Nature and Man: Essays Scientific and Philosophical* (London: 1888), pp. 284–315, on p. 293.
37 Carpenter, 'Ferrier's Experimental Investigations', p. 8.
38 Daniel Pick, *Faces of Degeneration: An European Disorder* (Cambridge: Cambridge University Press, 1989).
39 Thomas Richards, *The Imperial Archive: Knowledge and the Fantasy of Empire* (London: Verso, 1993).
40 Her parents did not hear of her death for two days. *Bell's Life in London and Sporting Chronicle*, 22 December 1883, p. 8, recorded that the marquis missed a day's attendance at the Victoria Derby and Melbourne Cup after receiving the news.

Part IV Mr Jeffery

13 Introducing Mr Jeffery

1 A Poor Victim, 'Electrical Belts', *Electrical Review*, 1887, 21: p. 648.
2 Lori Loeb, 'Consumerism and Commercial Electrotherapy: The Medical Battery Company in Nineteenth-Century London', *Journal of Victorian Culture*, 1999, 4: pp. 252–75.
3 'The Harness "Electropathic" Swindle', *Pall Mall Gazette*, 24 October 1893, 57: p. 1.
4 'The Medical Battery Company, Official Receiver's Report', *Pall Mall Gazette*, 3 January 1894, 58: p. 7.
5 Alfred Lewontin, *Dr G. Zander's Medico-Mechanical Gymnastics: Its Method, Importance and Application* (Stockholm: 1893).
6 Cornelius Bennett Harness, *Swedish Mechanical Exercises a Means of Cure and for the Prevention of Disease, with an Historical Introduction* (London: 1890).
7 'Official Receiver's Report', *Pall Mall Gazette*.
8 William Lynd, 'Curative Electricity', *Electrical Review*, 1888, 22: pp. 51–2.
9 'The "Electropathic and Zander Institute": A Protest', *Lancet*, 1889, I: pp. 190–1.
10 'Correspondence', *Electrical Review*, 1892, 31: pp. 242–3.
11 J.M.V. Money-Kent, 'Medical Electricity', *Electrician*, 1888, 20: p. 223.
12 Cornelius Bennett Harness, *A New Method of Treating Disease by Electricity* (London: 1883), p. 20.
13 Ibid.
14 Cornelius Bennett Harness, *A Treatise on the Special Diseases of Women and their Electropathic Treatment* (London: 1891), p. 5.
15 Ibid., p. 6.
16 Ibid., pp. 80, 36.

17 Edward Maitland, *Anna Kingsford: Her Life, Letters, Diary and Work* (London: 1896).

18 Carolyn Marvin, *When Old Technologies were New* (Oxford: Clarendon Press, 1980), pp. 109–51.

19 Iwan Rhys Morus, 'Currents from the Underworld: Electricity and the Technology of Display in early Victorian England', *Isis*, 1993, 84: pp. 50–69; Morus, *Frankenstein's Children*.

14 *Electric Entrepreneurs*

1   'Electricity in Medicine', *Electrician*, 1887, 20: p. 192.

2   'Electricity in Medicine', *Lancet*, 1895, II: pp. 587–8.

3   Henry Lewis Jones, 'Notes from the Electrical Department', *St Bartholomew's Hospital Reports*, 1892, 28: pp. 245–56.

4   William Edward Steavenson, 'The Electrical Department', *St Bartholomew's Hospital Reports*, 1883, 19: pp. 235–47.

5   Ibid., p. 236.

6   W.E. Steavenson, 'Report from the Electrical Department', *St Bartholomew's Hospital Reports*, 1886, 22: pp. 57–87.

7   W.S. Hedley, 'The Scope and Value of Electricity in Medicine', *Lancet*, 1896, I: pp. 1279–81.

8   'National Hospital for the Paralysed and Epileptic. The Electrical Room', *Lancet*, 1866, II: pp. 576–7.

9   W. Bolton Tomson, *Electricity in General Practice* (London: Baillere, Tindall & Cox, 1890), p. v.

10 K. Schall, *Illustrated Price List of Electro-medical Apparatus* (London: 1896), p. 57.

11 Herbert Tibbits, *How to use a Galvanic Battery in Medicine and Surgery: A Discourse upon Electro-therapeutics Delivered before the Hunterian Society upon November 8, 1876* (London: 1877), p. 1.

12 Herbert Tibbits, *Massage and its Applications: A Concluding Lecture Delivered to Nurses and Masseuses in connection with the West-end Hospital for Diseases of the Nervous System, Paralysis and Epilepsy* (London: 1887), p. 6.

13 Thomas Stretch Dowse, *The Modern Treatment of Disease by the System of Massage* (London: 1887), p. 92.

14 'New Companies Registered', *Electrical Review*, 1887, 20: p. 454.

15 Ibid., p. 477.

16 T.G. Hodgkinson, *The Electro-Neurotone Apparatus. Its Use & Application in Various Forms of Nervous and Other Diseases* (London: n.d.), p. ii.

17 William G. Johnson, *ElectroTherapie, the Art of Curing by Electricity* (London: 1886), p. 3.

18 E.S. D'Odiardi, *Medical Electricity. What is It? And How does It Cure?* (London: 1893), p. 57.

19 Advertisement for Harness' Electropathic Belts, pictured in Leonard de Vries, *Victorian Advertisements* (London: John Murray, 1968), p. 15. Originally from the *Illustrated London News*, 1890.

20 Advertisement for the Anti-Rheumatic Towel Co., pictured in ibid., p. 12. Originally from *Judy*, 1880.

21 Advertisement for Dr Scott's Electric Hair Brush in ibid., p. 10. Originally from *The Graphic*, 1883.

22 Michael Mason, *The Making of Victorian Sexuality* (Oxford: Oxford University Press, 1994), pp. 175–227.

23 It is not clear to what extent such views were symptomatic of the orthodox profession. See M. Jeanne Peterson, *The Medical Profession in mid-Victorian London* (Berkeley & Los Angeles CA: University of California Press, 1978). The popularity of works such as Acton's does tend to suggest that his ideas were influential amongst the lay population however.

24 Stewart, *Conservation of Energy* concludes his elementary survey of the conservation of energy with an account of its operation within the animal and human body. The impact of such views on medical matters, particularly with reference to female physical and psychiatric regulation, is discussed in Elaine Showalter, *The Female Malady: Women, Madness and English Culture 1830–1980* (London: Virago, 1985), pp. 124–6, and in Janet Oppenheim, *Shattered Nerves: Doctors, Patients and Depression in Victorian England* (Oxford: Oxford University Press, 1991), pp. 79–109. Stewart's exposition of the conservation of energy was very much the orthodox view, see Crosbie Smith, *The Science of Energy: A Cultural History of Energy Physics in Victorian Britain* (London: Athlone Press, 1998), pp. 247–55.

25 James Beresford Ryley, *Physical and Nervous Exhaustion in Man. Its Etiology and Treatment by 'Electro-Kinetics'* (London: 1892), p. 44.

26 Ibid., p. 60.

27 Ibid., p. 32.

28 James Beresford Ryley, *A Few Words of Advice to Young Wives* (London: 1895), p. 37.

29 See Daniel Pick, *Faces of Degeneration: A European Disorder, c.1848–c.1918* (Cambridge: Cambridge University Press, 1989).

30 James Beresford Ryley, *Electro-Magnetism and Massage in the Treatment of Rheumatic Gout, Dyspepsia, Sleeplessness, Nerve Prostration, and other Chronic Disorders* (London: 1885), p. 2.

31 Hodgkinson, *Electro-neurotone.*, p. 38.

32 Johnson, *Electro-therapie*, p. 7.

33 Hodgkinson, *Electro-neurotone*, p. 2.

34 Ibid., p. 17.

35 Joseph Granville Mortimer, 'Treatment of Pain by Mechanical Vibrations', *Lancet*, 1881, I: pp. 286–8.

36 Joseph Granville Mortimer, 'Nerve-vibration as a Therapeutic Agent', *Lancet*, 1882, I: pp. 949–51.

37 'Uterine Massage', *Lancet*, 1881, II: pp. 1094–5.

38 Joseph Granville Mortimer, *Nerves and Nerve Trouble* (London: 1884), p. 92.

39 Mortimer, *Nerves*, p. 57.

40 Rachel Maines, *The Technology of Orgasm* (Baltimore MD: Johns Hopkins University Press, 1999) characterises Mortimer's invention as a deliberate effort to mechanise the process of uterine massage. Maines is clearly wrong in this respect.

41 'Electricity and the Medical Profession', *Electrical Review*, 1882, 10: pp. 173–4.

42 William Henry Stone & W.J. Kilner, 'On Measurement in the Medical Application of Electricity', *Journal of the Society of Telegraph Engineers and of Electricians*, 1882, 11: pp. 107–28, on p. 109.

43 Takahiro Uyeama, 'Capital, Profession and Medical Technology: The Electro-therapeutic Institutes and the Royal College of Physicians, 1888–1922', *Medical History*, 1887, 41: pp. 150–81.

44 'The Institute of Medical Electricity', Undated pamphlet. Silvanus P. Thompson Collection, Institution of Electrical Engineers (SPT116/11).

45 Membership Certificates, IEE. MSS vol. 5A, p. 274 and MSS vol. 6B, p. 449.

46 'The Institute of Medical Electricity', Undated pamphlet. Silvanus P. Thompson Collection, Institution of Electrical Engineers (SPT116/11).

47 'A Visit to the Institute of Medical Electricity', *Electrician*, 1889, 23: pp. 490–2.

48 'The Electrified Room System', Undated pamphlet. Silvanus P. Thompson Collection, Institution of Electrical Engineers (SPT116/6).

49 A Visit to the Institute', *Electrician*, p. 490.

50 *Annals of the Royal College of Physicians*, 1888, 35: pp. 175–6. Meeting of Censor's Board, 10 April 1888.

51 Ibid., pp. 288–9.
52 Patricia Anderson, *The Printed Image and the Transformation of Popular Culture, 1790–1860* (Oxford: Clarendon Press, 1991), pp. 138–56.

15 *Measure for Measure*

1   Percy Pope, 'Five Cases of Lightning Stroke Occurring Simultaneously', *Lancet*, 1890, 136: pp. 718–9.
2   Simon Schaffer, 'Metrology, Metrication and Victorian Values', Bernard Lightman (ed.), *Victorian Science in Context* (Chicago IL: University of Chicago Press, 1997).
3   Wilhelm Erb, *Handbook of Electro-therapeutics*, translated by L. Putzel (New York: 1883), p. 19.
4   Armand de Watteville, *Practical Introduction to Medical Electricity*, 2nd edn (London: 1884), p. vi.
5   Armand de Watteville, 'On Current Measurements in Electrotherapeutics', *Lancet*, 1877, 110: p. 448.
6   Herbert Tibbits, 'On Current Measurements in Electrotherapeutics', *Lancet,* 1877, 110: p. 448.
7   William H. Stone & W.J. Kilner, 'On Measurement in the Medical Application of Electricity', *Journal of the Society of Telegraph Engineers and of Electricians*, 1882, 11: pp. 107–28, on p. 107.
8   'Electricity and the Medical Profession', *Electrical Review*, 1882, 10: pp. 173–4.
9   H. Newman Lawrence & A. Harries, 'Alternating v. Continuous Currents in Relation to the Human Body', *Journal of the Society of Telegraph Engineers and Electricians*, 1890, 19: pp. 290–316.
10 W.H. Stone, 'The Physiological Bearing of Electricity on Health', *Journal of the Society of Telegraph Engineers and Electricians*, 1884, 13: pp. 415–36, on p. 433.
11 J. Mount Bleyer, *The Best Method of Executing Criminals* (New York NY, 1888), pp. 429–31.
12 Ibid., pp. 431–2.
13 'Execution by Electricity', *British Medical Journal*, 1889, I: p. 35.
14 Mark Essig, *Edison and the Electric Chair* (Stroud: Sutton Publishing, 2003).
15 *New York Times*, 6 August 1890, p. 1.
16 *New York Times*, 7 August 1890, p. 1.
17 *The Times*, 7 August 1890, p. 3.
18 Ibid., p. 7.
19 'Capital Punishment by Electricity', *Lancet*, 1891, 138: pp. 943–4.
20 'Another Electrocution', *Lancet*, 1891, 138: p. 1349.
21 'Capital Punishment by Electricity', *Electrical Review*, 1891, 29: p. 495.
22 'The *Lancet* and Electrocution', *Electrical Review*, 1895, 36: p. 310.
23 S.F. Walker, 'Execution by Electricity', *Electrician*, 1889, 23: pp. 288–9.
24 Edward Tatum, 'Death from Electrical Currents', *New York Medical Times*, 1890, 51: pp. 207–9.
25 H. Newman Lawrence & Arthur Harries, 'Alternating versus Continuous Current in Relation to the Human Body', *Electrical Review*, 1890, 27: pp. 301–3.
26 'Dangers from Electric Shock', *Electrical Review*, 1890, 26: p. 379.
27 H. Newman Lawrence & Arthur Harries, 'Alternating versus Continuous Current in Relation to the Human Body', *Electrical Review*, 1890, 27: pp. 301–3.
28 'The Physiology of Death by Electric Shock', *Electrical Review*, 1895, 37: p. 530.
29 G.N. Stewart, 'Electricity: Alternating and Continuous Currents', *Lancet*, 1890, 135: p. 1040.
30 'Deaths and Accidents caused by High-Tension Currents', *Electrical Review*, 1893, 32: p. 374.
31 W.S. Hedley, 'The Pathology and Treatments of Electric Accidents', *Lancet*, 1894, II: p. 437.

32 'Electrocution', *Lancet*, 1895, I: p. 562.
33 H. Lewis Jones, 'The Lethal Effects of Electrical Currents', *British Medical Journal*, 1895, I: pp. 468–70.
34 Ibid., p. 469.
35 Thomas Oliver & Robert Bolam, 'On the Cause of Death by Electric Shock', *British Medical Journal*, 1898, I: pp. 132–5.
36 W.S. Hedley, 'The Pathology and Treatment of Electric Accidents', *Lancet*, 1894, 144: p. 437.
37 W.S. Hedley, 'The Painlessness of Electric Death', *Electrical Review*, 1897, 41: pp. 205–6.
38 W.S. Hedley, 'Apologia pro electricitate suâ', *Lancet*, 1895, 145: pp. 1105–9.
39 Benjamin Ward Richardson, 'Researches on the Treatment of Suspended Animation', *British & Foreign Medico-chirurgical Review*, 1863, 31: pp. 478–505, on p. 495.
40 Julius Althaus, 'Old Age and Rejuvenescence', *Lancet*, 1899, 153: pp. 149–52.
41 Arthur Conan Doyle, 'The Los Amigos Fiasco', *Round the Red Lamp, being Facts and Fancies of Medical Life* (London: 1894), pp. 218–94, on p. 282.
42 Ibid., p. 293.

16 Harness on Trial

1   'The Medical Battery Company v. Jeffery', *Electrical Review*, 1892, 31: pp. 99–101.
2   Ibid.
3   'Kissing the Book', *British Medical Journal*, 1901, I: p. 726. To swear an oath 'Scotch fashion' meant to swear whilst holding up one hand, rather than by touching or kissing the Bible.
4   'Electropathic Belts', *Electrical Review*, 1892, 31: pp. 101–2.
5   Ibid., p. 102.
6   'Electricity and the Medical Profession', *Electrical Review*, 1892, 31: pp. 365–7.
7   *Statement by Dr Tibbits*, printed pamphlet, 1892. Held by Royal College of Physicians. Strangely, at some stage in its history, this copy of the pamphlet has been torn in two by somebody and then repaired. Maybe some former fellow was expressing his contempt for the unfortunate Tibbits.
8   Herbert Tibbits, *The Treatment of Disease by the Application of Currents of Electricity of Low Power for Lengthened Periods* (London: 1892). Printed pamphlet held by Royal College of Physicians.
9   Ibid., p. 9.
10 Ibid., p. 5.
11 'Tibbits v. Alabaster and Others', *Electrical Review*, 1893, 32: pp. 184–8, 224–7; on p. 185.
12 Ibid., p. 186.
13 Ibid., p. 187.
14 Ibid., p. 224.
15 For Kelvin's massive reputation at the time see Crosbie Smith & M. Norton Wise, *Energy and Empire: A Biographical Study of Lord Kelvin* (Cambridge: Cambridge University Press, 1989).
16 'Tibbits v. Alabaster', p. 186.
17 'Alabaster and Others v. The Medical Battery Company', *Electrical Review*, 1893, 32: pp. 275–87, 733–5.
18 Just how much depends on how one calculates. Basing the calculation on the retail price index, £1,000 in 1893 would be worth about £80,000 in 2009. Basing the calculation on average earnings, probably a better indicator in this case, it would be closer to £500,000. A lot of money, in any case. See www.measuringworth.com/index, accessed on 20 May 2009.
19 'The Electropathic Belt Libel Suits', *Electrical Review*, 1893, 32: pp. 733–5.

20 'The Harness Electropathic Belt', *Electrical Review*, 1893, 33: p. 423.
21 The episode is discussed at some length in Judith Walkowitz, *City of Dreadful Delight: Narratives of Sexual Danger in Late Victorian London* (London: Virago, 1992), pp. 81–120.
22 'The Harness "Electropathic" Swindle', *Pall Mall Gazette*, 19 October 1893, p. 1.
23 Ibid.
24 'Correspondence', *Pall Mall Gazette*, 21 October 1893, p. 3.

Part V Back to the Future

17 *Back to the Future*

1 James Crichton-Browne, *The Doctor Remembers* (London: Duckworth, 1938), pp 195, 263.

Corsi, Pietro, *The Age of Lamarck* (Los Angeles & Berkeley CA: University of California Press, 1988).

Crosse, Andrew, 'Description of Some Experiments Made with the Voltaic Battery, by Andrew Crosse, Esq., of Broomfield, near Taunton, for the Purpose of Producing Crystals; in the Process of Which Experiments Certain Insects Constantly Appeared', *Annals of Electricity*, 1838, 2: pp. 246–57.

Crosse, Cornelia, *Memorials, Scientific and Literary, of Andrew Crosse, the Electrician* (London: 1857).

———, 'Science and Society in the Fifties', *Temple Bar*, 1891, 93: pp. 33–51.

———, *Red-letter Days of my Life* (London: 1892).

Cursory Observations upon the Lectures on Physiology, Zoology, and the Natural History of Man, Delivered in the Royal College of Surgeons by W. Lawrence, in a Series of Letters Addressed to that Gentleman, with a Concluding Letter to his Pupils, by One of the People Called Christians (London: 1819).

D'Odiardi, E.S., *Medical Electricity. What is It? And How does It Cure?* (London: 1893).

'Dangers from Electric Shock', *Electrical Review*, 1890, 26.

Darnton, Robert, *Mesmerism and the End of the Enlightenment in France* (Cambridge MA: Harvard University Press, 1967).

Daunton, Martin, *Poverty and Progress* (Oxford: Oxford University Press, 1995).

'Davenport's Electro-magnetic Engine', *Mechanic's Magazine*, 1837, 27: pp. 404–5.

Davy, Humphry, 'An Account of the Late Improvements in Galvanism … By John Aldini', *Edinburgh Review*, 1803, 3: pp. 194–8.

Davy, John, *Memoirs of the Life of Sir Humphry Davy* (London: 1836).

———, *Fragmentary Remains, Literary and Scientific, of Sir Humphry Davy, Bart.* (London: 1858).

'Deaths and Accidents caused by High-Tension Currents', *Electrical Review*, 1893, 32: p. 374.

Delbourgo, James, *A Most Amazing Scene of Wonders* (Cambridge MA: Harvard University Press, 2006).

'Description of Mr. Pepys' Large Galvanic Apparatus', *Tilloch's Philosophical Magazine*, 1803, 15: pp. 94–6.

Desmond, Adrian, *The Politics of Evolution* (Chicago IL: University of Chicago Press, 1989).

Ditton, Humphry, *The New Law of Fluids* (London: 1714).

Dowse, Thomas Stretch, *The Modern Treatment of Disease by the System of Massage* (London: 1887).

Doyle, Arthur Conan, 'The Los Amigos Fiasco', *Round the Red Lamp, being Facts and Fancies of Medical Life* (London: 1894), pp. 218–94.

Duchenne, Guilleme-Benjamin, *The Mechanism of Human Facial Expression*, edited and translated by R. Andrew Cuthbertson (Cambridge: Cambridge University Press, 1990).

Edwinsford, Wylke, *A Review of a Work, Entitled Remarks on Scepticism, by the Rev. T. Rennell* (London: 1819).

Eland, George, *The Lobb Family from the Sixteenth Century* (Oxford: Oxford University Press, 1955).

'Electricity and the Medical Profession', *Electrical Review*, 1882, 10: pp. 173–4.

'Electricity and the Medical Profession', *Electrical Review*, 1892, 31: pp. 365–7.

'Electricity in Medicine', *Electrician*, 1887, 20.

'Electricity in Medicine', *Lancet*, 1895, 147: pp. 587–8.

'Electrocution', *Lancet*, 1895, 146.

'Electropathic Belts', *Electrical Review*, 1892, 31: pp. 101–2.

Eliot, George, *The Mill on the Floss*, Great Writers edition (London: Marshall Cavendish, 1986).

Erb, Wilhelm, *Handbook of Electro-therapeutics*, translated by L. Putzel (New York: 1883).

Essig, Mark, *Edison and the Electric Chair* (Stroud: Sutton Publishing, 2003).

'Execution by Electricity', *British Medical Journal*, 1889, 97.

Failing, Anthony, *A History of the Maida Vale Hospital for Nervous Diseases* (London: Butterworth, 1958).

Fara, Patricia, *An Entertainment for Angels* (London: Icon Books, 2002).

Faraday, Michael, 'Fifteenth Series', *Experimental Researches in Electricity* (London: Taylor & Francis, 1839–52).

————, 'On Mental Education', *Experimental Researches in Chemistry and Physics* (London: 1859).

————, 'On Table-turning', *Experimental Researches in Chemistry and Physics* (London: 1859).

Farrar, W. V., 'Andrew Ure FRS and the Philosophy of Manufactures', *Notes & Records of the Royal Society*, 1973, 27: pp. 299–324.

Foote, George, 'Sir Humphry Davy and his Audience at the Royal Institution', *Isis*, 1952, 43: pp. 6–12.

Fox, Robert Were, 'Theoretical Views of the Origins of Mineral Veins', *Annals of Electricity*, 1838, 2: pp. 166–94.

'Galvanism', *Tilloch's Philosophical Magazine*, 1802, 14: pp. 191–2, 364–6.

Gassiot, John P., 'On Some Experiments made with Ruhmkorff's Induction Coil', *Philosophical Magazine*, 1854, 7: pp. 97–9.

Golinski, Jan, *Science as Public Culture* (Cambridge: Cambridge University Press, 1992).

Gooday, Graeme, *Domesticating Electricity: Technology, Uncertainty and Gender, 1880–1914* (London: Pickering & Chatto, 2008).

Grove, William Robert, *On the Correlation of Physical Forces* (London: London Institution, 1846).

————, 'An Address on the Importance of Physical Science in Medical Education', *British Medical Journal*, 1869, 57: pp. 485–7.

Gull, William W., 'A Further Report on the Value of Electricity as a Remedial Agent', *Guy's Hospital Reports*, 1852, 8: pp. 81–143.

Hall, Marshall, 'On the Functions of the Medulla Oblongata and Medulla Spinalis, and on the Excito-motory System of Nerves', *Proceedings of the Royal Society*, 1830–37, 3: pp. 463–4.

————, *Lectures on the Nervous System and its Diseases* (Philadelphia: 1836).

————, *Essays Chiefly on the Theory of the Paroxysmal Diseases of the Nervous System* (London: 1849).

Halse, William H., 'Wonderful Effects of Voltaic Electricity in Restoring Animal Life when the Sensorial Powers have Entirely Ceased or in other Words, when Death in the Common Acceptation of the Term has Actually Occurred', *Annals of Electricity*, 1840, 4: pp. 481–4.

————, *The Extraordinary Remedial Efficacy of Medical Galvanism, when Scientifically Administered* (London: n.d.).

Harness, Cornelius Bennett, *A New Method of Treating Disease by Electricity* (London: 1883).

————, *Swedish Mechanical Exercises a Means of Cure and for the Prevention of Disease, with an Historical Introduction* (London: 1890).

————, *A Treatise on the Special Diseases of Women and their Electropathic Treatment* (London: 1891).

'Harveian Society of London, February 19, 1863. Henry W. Fuller M.D., President, in the Chair. "On the Uses and Value of Galvanism and Electricity in General Practice" By Harry Lobb Esq.', *British Medical Journal*, 1863, 45: pp. 226–7.

'Harveian Society of London', *British Medical Journal*, 1874, 67.

Hawkins, Michael, 'A Great and Difficult Thing: Understanding and Explaining the Human Machine in Restoration England', in Iwan Rhys Morus (ed.), *Bodies/Machines* (Oxford: Berg, 2002), pp. 15–38.

Hay, Douglas, *Albion's Fatal Tree* (Harmondsworth: Penguin, 1988).

Hedley, W. S., 'The Pathology and Treatment of Electric Accidents', *Lancet*, 1894, 144.

————, 'Apologia pro electricitate suâ', *Lancet*, 1895, 145: pp. 1105–9.

Hedley, W.S. 'The Scope and Value of Electricity in Medicine', *Lancet*, 1896, 147: pp. 1279–81.
———, 'The Painlessness of Electric Death', *Electrical Review*, 1897, 41: pp. 205–6.
Heilbron, John, *Electricity in the 17th and 18th Centuries* (Berkeley & Los Angeles CA: University of California Press, 1979).
Herschel, John, *Preliminary Discourse on the Study of Natural Philosophy* (London: 1830).
Hilton, Boyd, *The Age of Atonement: The Influence of Evangelicalism on Social and Economic Thought, 1785–1865* (Oxford: Oxford University Press, 1992).
———, *A Mad, Bad and Dangerous People: England 1783–1846* (Oxford: Oxford University Press, 2006).
Hodgkinson, T.G., *The Electro-Neurotone Apparatus. Its Use & Application in Various Forms of Nervous and Other Diseases* (London: n.d.).
Holmes, Richard, *Coleridge: Early Visions* (London: Viking, 1990).
Hopkins, Harry, *The Long Affray* (London: Secker & Warburg, 1985).
Horn, Pamela, *The Rural World* (London: Hutchinson, 1980).
Hubbard, Geoffrey, *Cooke, Wheatstone and the Invention of the Electric Telegraph* (London: Routledge & Kegan Paul, 1965).
Hughes, H.M., 'Digest of One Hundred Cases of Chorea, Treated in the Hospital', *Guy's Hospital Reports*, 1846, 4: pp. 360–94.
Hunt, Bruce, 'The Ohm is where the Art is: British Telegraph Engineers and the Development of Electrical Standards', *Osiris*, 1994, 9: pp. 48–63.
Huxley, Thomas Henry, 'On the Hypothesis that Animals are Automata, and its History', *Fortnightly Review*, 1874, 16: pp. 555–80.
'Instinct and Reason', *British & Foreign Medico-chirurgical Review*, 1850, 6: pp. 522–4.
James, Frank, 'Michael Faraday, The City Philosophical Society and the Society of Arts', *Royal Society of Arts Journal*, 1992, 140: pp. 192–9.
——— (ed.), *The Correspondence of Michael Faraday*, vol. 3 (London: IEE, 1996).
Jay, Mike, *The Atmosphere of Heaven: The Unnatural Experiments of Dr Beddoes and his Sons of Genius* (London & New Haven: Yale University Press, 2009).
Jenkin, Fleeming, *Electricity and Magnetism* (London: 1873).
Johnson, Charles, *An Account of the Trials of the Ely and Littleport Rioters, in 1816* (Ely: C. Johnston, 1893).
Johnson, William G., *ElectroTherapie, the Art of Curing by Electricity* (London: 1886).
Jones, Henry Lewis, 'Notes from the Electrical Department', *St Bartholomew's Hospital Reports*, 1892, 28: pp. 245–56.
———, 'The Lethal Effects of Electrical Currents', *British Medical Journal*, 1895, 109: pp. 468–70.
King, Andrew & John Plunkett, *Victorian Print Media* (Oxford: Oxford University Press, 2005).
King, Peter, *Crime, Justice and Discretion in England, 1740–1820* (Oxford: Oxford University Press, 2000).
'Kissing the Book', *British Medical Journal*, 1901, 121.
Lawrence, H. Newman & Arthur Harries, 'Alternating versus Continuous Current in Relation to the Human Body', *Electrical Review*, 1890, 27: pp. 301–3.
Lawrence, H. Newman & A. Harries, 'Alternating v. Continuous Currents in Relation to the Human Body', *Journal of the Society of Telegraph Engineers and Electricians*, 1890, 19: pp. 290–316.
Lawrence, William, *An Introduction to Comparative Anatomy and Physiology* (London: 1816).
———, *Lectures on Physiology, Zoology, and the Natural History of Man* (London: 1819).
Laycock, Thomas, *An Essay on Hysteria* (Philadelphia: 1840).
———, *A Treatise on the Nervous Diseases of Women* (London: 1840).
———, 'On the Treatment of Cerebral Hysteria, and of Moral Insanity in Women, by Electro-galvanism', *Medical Times*, 1850, 1: pp. 57–8.

Laycock, Thomas, 'Woman in her Psychological Relations', *Journal of Psychological Medicine & Mental Pathology*, 1851, 4: pp. 18–50.

Leithead, William, *Electricity; Its Nature, Operation and Importance in the Phenomena of the Universe* (London: 1837).

Levere, Trevor, *Poetry Realised in Nature* (Cambridge: Cambridge University Press, 1981).

Lewontin, Alfred, *Dr G. Zander's Medico-Mechanical Gymnastics: Its Method, Importance and Application* (Stockholm: 1893).

Lightman, Bernard, *Victorian Popularisers of Science: Designing Nature for New Audiences* (Chicago: University of Chicago Press, 2007).

Lobb, Harry, *On Some of the More Obscure Forms of Nervous Affections: Their Pathology and Treatment* (London: 1858).

———, *On the Curative Treatment of Paralysis and Neuralgia, and Other Affections of the Nervous System with the Aid of Galvanism* (London, 1859).

———, 'Acute Neuralgia Cured with the Aid of the Continuous Current, with Peculiar Sympathetic Effect upon the Uterus', *Lancet*, 1860, 76.

———, *On Some of the More Obscure Forms of Nervous Affections: Their Pathology and Treatment*, 2nd edn, (London: 1863).

———, 'Special Hospitals – the London Galvanic Hospital', *Lancet*, 1863, 82.

———, 'An Attempt to Adapt the Laws of Electro-dynamics to the Science of Medical Electricity', *British & Foreign Medico-chirurgical Review*, 1866, 73: pp. 508–16.

———, *A Popular Treatise on Curative Electricity, Especially Addressed to Sufferers from Paralysis, Rheumatism, Neuralgia, and Loss of Nervous & Physical Power* (London: 1867).

Loeb, Lori, 'Consumerism and Commercial Electrotherapy: The Medical Battery Company in Nineteenth-Century London', *Journal of Victorian Culture*, 1999, 4: pp. 252–75.

Lynd, William, 'Curative Electricity', *Electrical Review*, 1888, 22: pp. 51–2.

Macceroni, Francis, 'An Account of Some Remarkable Phenomena seen in the Mediterranean, with Some Physiological Deductions', *Mechanics' Magazine*, 1831, 15: pp. 93–6.

———, *Memoirs and Adventures of Colonel Maceroni, Late Aide-de-Camp to Joachim Murat, King of Naples* (London: 1838).

Macilwain, George, *Memoirs of John Abernethy* (London: 1853).

Mackenzie, Peter, *Old Reminiscences of Glasgow and the West of Scotland*, 2 vols, (Glasgow: 1890).

Mackintosh, Thomas Simmons, *The Electrical Theory of the Universe* (Boston: 1846).

Maines, Rachel, *The Technology of Orgasm* (Baltimore MD: Johns Hopkins University Press, 1999).

Maitland, Edward, *Anna Kingsford: Her Life, Letters, Diary and Work* (London: 1896).

Marvin, Carolyn, *When Old Technologies were New* (Oxford: Clarendon Press, 1980).

Marx, Karl, *Capital*, vol. 1 (Harmondsworth: Penguin Books, 1982).

Mason, Michael, *The Making of Victorian Sexuality* (Oxford: Oxford University Press, 1994).

Maudsley, Henry, 'Sex in Mind and Education', *Fortnightly Review*, 1874, 15: pp. 479–80.

Maxwell, James Clerk, 'Electricity and Magnetism', *Nature*, 15 May 1873, pp. 42–3.

McCalman, Iain, *Radical Underworld* (Cambridge: Cambridge University Press, 1988).

'Medical Advertisements in the Daily Journals', *British Medical Journal*, 1874, 67.

Milligan, J.G., *Mind and Matter* (London: 1847).

Mingay, G.E., *Mrs Hurst Dancing, and Other Scenes from Regency Life* (London: Gollancz, 1981).

Mitchell, Leslie, *Bulwer Lytton: The Rise and Fall of a Victorian Man of Letters* (London: Hambledon & London, 2003).

'Modern Scepticism, as Connected with Organisation and Life', *British Review and London Critical Journal*, 1819, 14: pp. 169–205.

Money-Kent, J.M.V., 'Medical Electricity', *Electrician*, 1888, 20.

Morrell, Jack & Arnold Thackray, *Gentlemen of Science: The Early Years of the British Association for the Advancement of Science* (Oxford: Oxford University Press, 1981).

Mortimer, Joseph Granville, 'Treatment of Pain by Mechanical Vibrations', *Lancet*, 1881, 118: pp. 286–8.

——, 'Nerve-vibration as a Therapeutic Agent', *Lancet*, 1882, 120: pp. 949–51.

——, *Nerves and Nerve Trouble* (London: 1884).

Morus, Iwan Rhys, 'Batteries, Bodies & Belts: Making Careers in Victorian Medical Electricity', in Paola Bertucci and Giuliano Pancaldi (eds), *Electric Bodies: Episodes in the History of Medical Electricity* (Bologna: Universita di Bologna, 2001), pp. 209–38.

——, 'Correlation and Control: William Robert Grove and the Construction of a New Philosophy of Scientific Reform', *Studies in History & Philosophy of Science*, 1991, 22: pp. 589–621.

——, 'Currents from the Underworld: Electricity and the Technology of Display in early Victorian England', *Isis*, 1993, 84: pp. 50–69.

——, 'The Electric Ariel: Telegraphy and Commercial Culture in early Victorian England', *Victorian Studies*, 1996, 29: pp. 339–78.

——, *Frankenstein's Children: Electricity, Exhibition and Experiment in early Nineteenth-century London* (Princeton: Princeton University Press, 1998).

——, *Michael Faraday and the Electrical Century* (London: Icon Books, 2004).

——, 'More the Aspect of Magic than Anything Natural: The Philosophy of Demonstration', in Bernard Lightman & Aileen Fyfe (eds), *Science in the Marketplace: Nineteenth-century Sites and Experiences* (Chicago: University of Chicago Press, 2007), pp. 336–70.

Murphy, Michael, *Cambridge Newspapers and Opinion, 1780–1850* (Cambridge: Oleander Press, 1977).

Murphy, Patrick, *Rudiments of the Primary Forces of Gravity, Magnetism, and Electricity, in their Agency on the Heavenly Bodies* (London: 1830).

Murray, James, *Electricity as a Cause of Cholera, or Other Epidemics, and the Relation of Galvanism to the Action of Remedies* (Dublin: 1849).

'National Hospital for the Paralysed and Epileptic. The Electrical Room', *Lancet*, 1866, 89: pp. 576–7.

Nead, Lynda, *Victorian Babylon* (London: Yale University Press, 2005).

Newton, Isaac, *Opticks*, 4th edn, (London: 1730).

Noad, Henry M., *A Course of Eight Lectures on Electricity, Galvanism, Magnetism, and Electro-magnetism* (London: 1844).

——, *The Inductorium, or Induction Coil* (London: 1868).

Nunn, Thomas William, *Inflammation of the Breast, and Milk Abscess* (London: 1853).

Okes, Thomas Verney, *An Account of the Providential Preservation of Elizabeth Woodstock, who Survived a Confinement Under the Snow of nearly Eight Days and Nights* (Cambridge: 1799).

Oliver, Thomas & Robert Bolam, 'On the Cause of Death by Electric Shock', *British Medical Journal*, 1898, 115: pp. 132–5.

Oppenheim, Janet, *Shattered Nerves: Doctors, Patients and Depression in Victorian England* (Oxford: Oxford University Press, 1991).

Pepper, John Henry, *The Boy's Playbook of Science* (London: 1866).

——, *The True History of the Ghost; and all about Metempsychosis* (London: Cassell & Co., 1890).

Pera, Marcello, *The Ambiguous Frog* (Princeton NJ: Princeton University Press, 1992).

Peterson, M. Jeanne, *The Medical Profession in mid-Victorian London* (Berkeley & Los Angeles CA: University of California Press, 1978).

Phillips, Richard, 'A Brief Account of a Visit to Andrew Crosse, Esq., of Broomfield', *Annals of Electricity*, 1836–37, 1: pp. 135–45.

Pick, Daniel, *Faces of Degeneration: An European Disorder* (Cambridge: Cambridge University Press, 1989).

Piggot, W.P., *Galvanic Belts, and Galvanism* (London: 1852).

Pine, Thomas, 'On the Probable Connection between Electricity and Vegetation', *Mechanics' Magazine*, 1835–36, 24: pp. 99–104.

Pollock, Lady Jane (writing anonymously), 'Michael Faraday', *St Paul's Magazine*, 1870, 4: pp. 293–303.

Pope, Percy, 'Five Cases of Lightning Stroke Occurring Simultaneously', *Lancet*, 1890, 136: pp. 718–9.

Porter, Roy, *Health for Sale* (Manchester: Manchester University Press, 1989).

———, *London: A Social History* (London: Hamish Hamilton, 1995).

———, *Enlightenment* (Harmondsworth: Penguin, 2001).

———, *Flesh in the Age of Reason* (London: Allen Lane, 2003).

Practical Guide for the Electro-medical Treatment of Rheumatic and Nervous Diseases, Gout, etc., by Means of Pulvermacher's Hydro-electric Chains (London: Pulvermacher's General Depot, 1856).

Prescott, G.B., *History, Theory and Practice of the Electric Telegraph* (Boston: 1860).

Priestley, Joseph, *History and Present State of Electricity* (London, 1767).

———, *Experiments and Observations on Different Kinds of Air* (Birmingham: Thomas Pearson, 1790).

'Recent Controversy on Materialism', *Monthly Repository of Theology and General Literature*, 182, 17: pp. 170–82.

Rennell, Thomas, *Remarks on Scepticism, Especialy as it is Connected with the Subjects of Organisation and Life, Being an Answer to the Views of M. Bichat, Sir T.C. Morgan, and Mr. Lawence* (London: 1819).

'Report Presented to the Class of the Exact Sciences of the Academy of Turin, 15 August 1802, in Regard to the Galvanic Experiments made by C. Vassali-Eandi, Giulio, and Rossi, on the 10 and 14 of the Same Month, on the Head and Trunk of Three Men a Short Time after their Decapitation', *Tilloch's Philosophical Magazine*, 1803, 15: pp. 38–45.

'Reviews and Notices Respecting New Books', *Philosophical Magazine*, 1838, 11: pp. 127–30.

Richards, Thomas, *The Imperial Archive: Knowledge and the Fantasy of Empire* (London: Verso, 1993).

Richardson, Benjamin Ward, 'Researches on the Treatment of Suspended Animation', *British & Foreign Medico-chirurgical Review*, 1863, 31: pp. 478–505.

———, 'On Research with the Large Induction Coil of the Royal Polytechnic Institution, with Special Reference to the Cause and Phenomena of Death by Lightning', *Medical Times and Gazette*, 1869, 38: pp. 39, 183–6, 511–4, 595–9.

———, *Hygeia: The City of Health* (London: 1876).

Richardson, Ruth, *Death, Dissection and the Destitute* (London: Routledge & Kegan Paul, 1987).

Rive, Auguste de la, 'Some Notes on the Present State of the Study of Electricity in England, Collected During a Recent Sojourn in that Country', *Electrical Magazine*, 1845, 1: pp. 100–7.

Rook, Arthur, Margaret Carlton & W. Graham Cannon, *The History of Addenbrooke's Hospital, Cambridge* (Cambridge: Cambridge University Press, 1992).

Ross, Sidney, 'Scientist: The Story of a Word', *Annals of Science*, 1962, 18: pp. 65–85.

Rule, John, *Albion's People* (London: Longman, 1992).

Russet, Cynthia Eagle, *Sexual Science: The Victorian Construction of Womanhood* (Cambridge MA: Harvard University Press, 1981).

Ruston, Sharon, *Shelley and Vitality* (London: Palgrave Macmillan, 2005).

Rutter, J.O.N., *Human Electricity: The Means of its Development, Illustrated by Experiments* (London: 1854).

Ryley, James Beresford, *Electro-Magnetism and Massage in the Treatment of Rheumatic Gout, Dyspepsia, Sleeplessness, Nerve Prostration, and other Chronic Disorders* (London: 1885).

———, *Physical and Nervous Exhaustion in Man. Its Etiology and Treatment by 'Electro-Kinetics'* (London: 1892).

———, *A Few Words of Advice to Young Wives* (London: 1895).

Schaffer, Simon, 'Metrology, Metrication and Victorian Values', Bernard Lightman (ed.), *Victorian Science in Context* (Chicago IL: University of Chicago Press, 1997).

———, 'Enlightened Automata', in William Clark, Jan Golinski & Simon Schaffer (eds), *The Sciences in Enlightened Europe* (Chicago IL: University of Chicago Press, 1999).

Schall, K., *Illustrated Price List of Electro-medical Apparatus* (London: 1896).

Searby, Peter, *A History of the University of Cambridge*, vol. 2, (Cambridge: Cambridge University Press, 1997).

Secord, James A., 'Extraordinary Experiment: Electricity and the Creation of Life in Victorian England', in David Gooding, Trevor Pinch & Simon Schaffer (eds), *The Uses of Experiment* (Cambridge: Cambridge University Press, 1989), pp. 337–83.

——— (ed.), *Vestiges of the Natural History of Creation and Other Evolutionary Writings by Robert Chambers* (Chicago IL: University of Chicago Press, 1994).

———, *Victorian Sensation: The Extraordinary Publication, Reception and Secret Authorship of Vestiges of the Natural History of Creation* (Chicago: University of Chicago Press, 2000).

Sharples, Eliza, 'An Inquiry How Far the Human Character is Formed by Education or External Circumstances', *The Isis*, 1832, 1: pp. 81–5.

———, 'Fifth Discourse on the Bible', *The Isis*, 1832, 1: pp. 241–7.

Showalter, Elaine, *The Female Malady: Women, Madness and English Culture 1830–1980* (London: Virago, 1985).

Smee, Alfred, *Elements of Electrometallurgy*, 3rd edn, (London: 1844).

———, *Elements of Electrobiology* (London: 1848).

———, *Instinct and Reason, Deduced from Electro-biology* (London: 1850).

Smith, Crosbie & M. Norton Wise, *Energy and Empire: A Biographical Study of Lord Kelvin* (Cambridge: Cambridge University Press, 1989).

———, *The Science of Energy: A Cultural History of Energy in Victorian Britain* (London: Athlone, 1998).

Snape, Joseph, *Electro-Dentistry: Facts, and Observations* (London: 1865).

'Spirit of Discovery', *Mirror of Literature, Amusement and Instruction*, 1830, 15: pp. 95–6.

Stallybrass, Oliver, 'How Faraday Produced Living Animalculae: Andrew Crosse and the Story of a Myth', *Proceedings of the Royal Institution*, 1967, 41: pp. 597–619.

Stanton, Henry B., *Sketches of Reforms and Reformers of Great Britain and Ireland* (New York: 1849).

Steavenson, William Edward, 'The Electrical Department', *St Bartholomew's Hospital Reports*, 1883, 19: pp. 235–47.

———, 'Report from the Electrical Department', *St Bartholomew's Hospital Reports*, 1886, 22: pp. 57–87.

Stein, Dorothy, *Ada: A Life and a Legacy* (Cambridge MA: MIT Press, 1985).

Stewart, Balfour, *The Conservation of Energy: Being an Elementary Treatise on Energy and its Laws* (London: 1873).

Stewart, G.N., 'Electricity: Alternating and Continuous Currents', *Lancet*, 1890, 135.

Stone, William H., 'The Physiological Bearing of Electricity on Health', *Journal of the Society of Telegraph Engineers and Electricians*, 1884, 13: pp. 415–36.

Stone, William H. & W.J. Kilner, 'On Measurement in the Medical Application of Electricity', *Journal of the Society of Telegraph Engineers and of Electricians*, 1882, 11: pp. 107–28.

Strutt, Robert, *The Life of Lord Rayleigh* (London: Arnold, 1924).

Sturgeon, William, 'On Electro-magnetism', *Philosophical Magazine*, 1824, 64: pp. 242–9.

———, 'A General Outline of the Various Theories which have been Advanced for the Explanation of Terrestrial Magnetism', *Annals of Electricity*, 1836–37, 1: pp. 117–23.

———, *A Course of Twelve Elementary Lectures on Galvanism* (London: 1843).

Sutton, Geoffrey, 'The Politics of Science in early Napoleonic France: the Case of the Voltaic Pile', *Historical Studies in the Physical Sciences*, 1981, 11: pp. 329–66.

Tatum, Edward, 'Death from Electrical Currents', *New York Medical Times*, 1890, 51: pp. 207–9.

'The "Electropathic and Zander Institute": A Protest', *Lancet*, 1889, 134: pp. 190–1.

'The Electropathic Belt Libel Suits', *Electrical Review*, 1893, 32: pp. 733–5.

'The Harness Electropathic Belt', *Electrical Review*, 1893, 33.

'The *Lancet* and Electrocution', *Electrical Review*, 1895, 36.

'The March of Specialism', *Lancet*, 1863, 82.

'The Medical Battery Company v. Jeffery', *Electrical Review*, 1892, 31: pp. 99–101.

'The Physiology of Death by Electric Shock', *Electrical Review*, 1895, 37.

'The Present State of the Science of Medical Electricity', *Electrician*, 1863, 5.

The Radical Triumvirate, or, Infidel Paine, Lord Byron, and Surgeon Lawrence, Colleaguing with the Patriotic Radicals to Emancipate Mankind from all Laws Human and Divine (London: 1820).

The Works of the Right Honourable Edmund Burke (London: 1834).

Thelwall, John, *The Rights of Nature Against the Usurpations of the Establishments* (Norwich: 1796).

Thomson, Ann (ed.), *Machine Man and Other Writings* (Cambridge: Cambridge University Press, 1996).

'Tibbits v. Alabaster and Others', *Electrical Review*, 1893, 32: pp. 184–8, 224–7.

Tibbits, Herbert, *Handbook of Medical and Surgical Electricity*, 2nd edn, (London: J.A. Churchill, 1877).

———, *How to use a Galvanic Battery in Medicine and Surgery: A Discourse upon Electro-therapeutics Delivered before the Hunterian Society upon November 8, 1876* (London: 1877).

———, 'On Current Measurements in Electrotherapeutics', *Lancet*, 1877, 110.

———, *Massage and its Applications: A Concluding Lecture Delivered to Nurses and Masseuses in connection with the West-end Hospital for Diseases of the Nervous System, Paralysis and Epilepsy* (London: 1887).

———, *The Treatment of Disease by the Application of Currents of Electricity of Low Power for Lengthened Periods* (London: 1892).

Tilt, Edward John, *On Diseases of Women and Ovarian Inflammation* (London: 1853).

Tomson, W. Bolton, *Electricity in General Practice* (London: 1890).

Toole, Alexandra (ed.), *Ada, The Enchantress of Numbers* (Mill Valley CA: Strawberry Press, 1992).

Tyndall, John, *Faraday as a Discoverer* (London: 1868)

———, *Address Delivered before the British Association Assembled at Belfast, with Additions* (London: 1874).

Ure, Andrew, 'An Account of Some Experiments made on the Body of a Criminal immediately after Execution, with Physiological and Practical Observations', *Quarterly Journal of Science*, 1819, 6: pp. 283–94.

———, *The Philosophy of Manufactures* (London: 1835).

'Uterine Massage', *Lancet*, 1881, 119: pp. 1094–5.

Uyeama, Takahiro, 'Capital, Profession and Medical Technology: The Electro-therapeutic Institutes and the Royal College of Physicians, 1888–1922', *Medical History*, 1887, 41: pp. 150–81.

von Haller, Albrecht, 'A Dissertation on the Sensible and Irritable Parts of Animals', *Bulletin of the History of Medicine and Allied Sciences*, 1936, 4: pp. 651–99.

Vries, Leonard de, *Victorian Advertisements* (London: John Murray, 1968).

Walker, Charles V., 'Mr. Gassiot's Electrical Soirée', *Electrical Magazine*, 1845, 1.

——, 'Facts [?] in the History of Electricity', *Electrical Magazine*, 1845, 1: pp. 551–2.

——, 'Notices of New Books', *Electrical Magazine*, 1845, 1: pp. 542–50.

Walker, S.F., 'Execution by Electricity', *Electrician*, 1889, 23: pp. 288–9.

Walkowitz, Judith, *City of Dreadful Delight: Narratives of Sexual Danger in Late Victorian London* (London: Virago, 1992).

Warwick, Andrew, *Masters of Theory* (Chicago: University of Chicago Press, 2003).

Watteville, Armand de, 'On Current Measurements in Electrotherapeutics', *Lancet*, 1877, 110.

——, *Practical Introduction to Medical Electricity*, 2nd edn, (London: 1884).

Weekes, William H., *Proceedings of the London Electrical Society*, 1841.

Wesley, John, *The Desideratum, or Electricity made Plain and Useful* (London: 1760).

Wetzels, Walter, 'Aspects of Natural Science in German Romanticism', *Studies in Romanticism*, 1971, 10: pp. 44–59.

——, 'Johann Wilhelm Ritter: Romantic Physics in Germany', in Andrew Cunningham & Nick Jardine (eds), *Romanticism and the Sciences* (Cambridge: Cambridge University Press, 1990).

Wheeler, Stephen, *The Poetical Works of Walter Savage Landor* (Oxford: Clarendon Press, 1937).

Whewell, William, *The Philosophy of the Inductive Sciences* (London: 1840).

Wilkinson, Charles Henry, *An Essay on the Leyden Phial, with a View to Explaining this Remarkable Phenomenon in Pure Mechanical Principles* (London: 1798).

——, *Elements of Galvanism, in Theory and Practice* (London: 1804).

Wilks, S. & G.T. Bettany, *A Biographical Dictionary of Guy's Hospital* (London: 1892).

Williams, L. Pearce, *Michael Faraday* (London: Chapman & Hall, 1965).

—— (ed.), *The Selected Correspondence of Michael Faraday* (Cambridge: Cambridge University Press, 1971).

Williams, Thomas, 'On the Laws of the Nervous Force, and the Function of the Roots of the Spinal Nerve', *Lancet*, 1847, 51: pp. 516–7.

Wilson, George, *Electricity and the Electric Telegraph* (London: 1855).

Winter, Alison, 'Harriet Martineau and the Reform of the Invalid', *Historical Journal*, 1995, 38: pp. 597–616.

——, 'A Calculus of Suffering: Ada Lovelace and the Bodily Constraints on Women's Knowledge in early Victorian England', in Christopher Lawrence & Steven Shapin (eds), *Science Incarnate* (Chicago: University of Chicago Press, 1998), pp. 202–39.

——, *Mesmerized* (Chicago: University of Chicago Press, 1998).

Wynter, Andrew, 'The Electric Telegraph', *Quarterly Review*, 1854, 95.

Yatman, Matthew, *Galvanism, Proved to be a Regular Assistant Branch of Medicine* (London: 1810).

INDEX

Abernethy, John 40–8, 49, 55
Acton, William 152
Adelaide Gallery 72–5, 94
Aldini, Giovanni 24–7, 29, 31, 35–8, 43, 48, 50, 55–7, 81, 136, 169
Althaus, Julius 109–10, 125–7, 134, 170
Analytical Engine 62, 100
animal electricity 23, 25, 29–30
Armstrong, William George 72, 114–5
Association of Medical Electricians 140, 175
atmospheric electricity 59, 64, 67, 76, 80
automatism 132–4

Babbage, Charles 58, 62–3, 68, 74, 88, 92, 94, 98–100
Banks, Sir Joseph 30, 56, 99
baths, electric 10, 83, 86, 108, 138, 146, 156
battle of the systems 163–4, 166
Beddoes, Thomas 25, 31, 33
Bellhouse, Dawson 85–6
Bird, Golding 83–6, 91, 105, 146
    magneto–electric machine 73, 83, 148
bodily electricity 10, 54, 69, 88, 110, 121, 144, 168, 176, 182
body, human 21–2, 25, 28, 30, 32, 42, 45, 78, 87, 96, 114, 129–30, 144, 152, 154, 160–3, 166–9, 185
British Association for the Advancement of Science 63, 65, 77, 98, 131–2, 167
Brown, Harold 163–4, 166
Buckland, William 65, 67
Bulwer Lytton, Edward 112–4, 117, 121, 123, 135
    Coming Race, The 112–3, 121, 135
Burke, Edmund 27, 39–40
Burrough, Justice 12, 15
Byron, George Gordon 8, 19, 38, 45, 58–62, 94, 100

Calculating Engine 62, 99
Cambridge 11–9, 46, 49, 53–4
    University of 13, 16–8, 45–6, 53, 58, 61–2, 74, 99, 121–3, 135, 146
Carlile, Richard 47, 72, 82, 85
Carpenter, William Benjamin 128–9, 132–4, 155
Carpue, Joseph 25, 34–5, 38
Carter, Robert Brudenell 89–90
Cavendish Laboratory 121–2, 135
Chambers, Robert 79–80
Clark, Latimer 162
Clydesdale, Matthew 50–6
Coleridge, Samuel Taylor 28, 33–4, 37, 39, 47–8, 64, 179
Combe, George 98
conservation of energy 130–1, 152
Cooke, William Fothergill 70–1
correlation of physical forces 120, 128–30
Cragside 114
Crichton–Browne, James 183
Crookes, William 121, 133
Crosse, Andrew 58–9, 61–8, 72–5, 77–8, 80, 81, 91, 93, 97, 100–1
    acarus crossii 67, 80
Cumming, James 16–8, 50, 53–7
Cuthbertson, John 25–6, 30, 35
consumerism 7, 9, 141, 184

D'Arsonval, Arsene 168–9
Davy, Sir Humphry 8, 25, 28, 30–4, 42, 47, 56–7, 64, 70–1, 75, 94
de la Mettrie, Julien Offrey 27, 40
de Morgan, Augustus 58, 60
de Watteville, Armand 146, 161, 171, 178
degeneration 134, 152–3, 185
Dickens, Charles 65, 113
discipline 9, 61–2, 88, 94–6, 98–101, 111, 122–3, 125, 127, 134–5, 160, 171, 182, 185
dissection 15–6, 18, 25–6, 35, 50–1, 62

Divan, William 54
Duchenne, Guilleme Benjamin 126–7

Edison, Thomas Alva 163–4, 166
Edwinsford, Wylke 46–7
electric chair 10, 168
electric eels 73, 75, 94
electric locomotives 70, 73, 115
electric shocks 21–2, 30, 63–4, 91, 114, 118,
    122, 162, 167–8, 170, 184
electric sparks 22, 30, 72, 77, 84, 114, 120,
    122, 184
electrical experimentation 21, 23, 30, 49,
    64–6, 75, 78, 101, 141, 184–5
Electrical Review, The 137, 139, 150, 165–6,
    167, 174–80
Electrician, The 106, 140, 145–6, 156, 166
electrocution 159–60, 165–70
electromagnetic induction 70, 74, 83–4
electromagnetic machine 73, 115, 126
Electropathic and Zander Institute 139–41,
    179; *see also* Medical Battery Company
electropathic belts 139–43, 146, 171, 172,
    176–7, 179–82
electrotherapy 84, 103, 125, 127, 131, 145,
    148–50
Elliotson, John 97
enlightenment 8, 20, 22, 30, 32, 40, 50, 64,
    96
Erb, Wilhelm 139, 160

Faraday, Michael 9, 32, 39, 56–7, 59, 67–8,
    70–6, 80, 83–4, 93–7, 100, 130, 184
Ferrier, David 134
FitzGerald, Desmond 106, 140
Forster, George 25–7, 31, 35, 49, 55–6
Fox, Robert Were 77–8
Frankenstein 37–8, 49
Fyne Court 58–9, 64, 66–7, 70, 72, 75, 80,
    92, 97, 100–1

Galvani, Luigi 23, 27, 29
galvanic batteries 16–8, 31–2, 53, 75, 78–9,
    83, 89, 108, 126, 129, 141
galvanic belts 36, 81, 86, 153; *see also*
    electropathic belts
galvanic medicine 36, 82–91
galvanic rings 86, 88, 136
Gassiot, John Peter 73–4, 119–21
    Gassiot's cascade 119–21
Gatehouse, G.E. 174–5

gentlemen of science 63, 65–8, 75, 79, 88, 95,
    97–8, 101, 184
God 10, 12, 21, 28, 40, 44, 46, 80, 85, 94–5,
    99–100, 131
Godwin, William 11, 37
Graham, James 34, 152
Great Exhibition 115–6
Grieg, Woronzow 61, 92
Grove, William Robert 75, 120–1, 128–30
Guy's Hospital 35, 82–4, 105, 146

Hall, Marshall 82–3, 88, 128
Halse, William Hooper 84–6, 91
Harness, Cornelius Bennett 137–44, 145–6,
    153, 157, 159, 171, 172–82
Harries, Arthur 155, 162, 166–7, 178
Hartley, David 32, 42
Hatfield 115
Hedley, William Snowden 147, 149, 168–9
Herschel, John 62–3, 88, 99
Hodgkinson, T.G. 150–1, 153
Hospital for Sick Children 109, 148–9
Hunter, John 24, 41–2, 44–5
Huxley, Thomas Henry 132–3
Hydro-electric machine 72–3
hysteria 88–91, 102, 125, 127, 142, 152, 154

illumination, electrical 116, 121
induction coil 83–4, 118–20, 123, 148, 156, 168
    Great Induction Coil 118–9
Institute of Electrical Engineers 155, 168; *see
    also* Society of Telegraph Engineers
Institute of Medical Electricity 155–7, 162,
    178
International Exhibition (1862) 112, 115

Jeffery, Mr D. 9–10, 136–46, 149, 157, 159,
    170–2
Jeffrey, James 50, 54
Jenkin, Fleeming 122
Johnson, William 151, 153
Jones, Henry Lewis 146, 168, 178

Kemmler, William 159, 164–6
King, William 61
Kingsford, Anna 142–3

Lamarck, Jean Baptiste 40
Lancet, The 41, 55, 82, 110–1, 140, 145–6,
    154, 161, 165–7, 172–3
Lardner, Dionysius 97

Lawrence, Henry Newman 155–7, 162, 166–7
Lawrence, William 43–8, 49, 55
Laycock, Thomas 89–91, 102, 124, 128
Leithead, William 73, 79, 87, 91
life, nature of 20, 24–5, 27, 29–30, 31, 37–8,
    40, 41–8, 66–7, 75–6, 78–9, 130–1, 169–70
light, electric 70, 115–7, 123, 135, 168
Lobb, Harry 102–10, 119, 124–7, 129–30,
    134, 150
Locock, Charles 61, 92, 97, 104–7
London Electrical Society 73–4, 76–7, 79,
    83–5, 100
London Galvanic Hospital 106, 110, 119, 127
London Institution 63, 75
Lovelace, Ada 8–9, 58–69, 70, 72, 74, 79–80,
    91, 92–101, 104, 184
Lynd, William 139–40

Maceroni, Francis 77–9
Mackintosh, Thomas Simmons 72, 78–80
magic lanterns 117–8
Martineau, Harriet 97–8, 100
masturbation 89–90, 127
materialism 19, 27–8, 36, 38, 46–7, 56, 71, 75,
    81, 91, 93, 128, 131, 184
mathematics 58, 60–1, 68
Matteucci, Carlo 74, 82, 86
Maudsley, Henry 131
Maxwell, James Clerk 121–2, 130
measurement 9–10, 122, 146, 155, 160–1; see
    also metrology
Medical Battery Company 136–41, 143–4,
    153, 172–5, 179, 181–2, 184; see also
    Electropathic and Zander Institute
medical electricity 34, 81, 86, 102, 105–6,
    109–11, 125, 129–31, 140, 143, 145–8,
    150, 155, 157, 160–2, 177–8
mental education 50, 95–6, 98
mesmerism 83, 96–8, 100, 113
metrology 160, 171, 182; see also
    measurement
Mill, John Stuart 133
Money–Kent, J.M.V. 140–1
Mortimer, Joseph Granville 153–4
Murray, James 87

Napoleon Bonaparte 27, 31, 56, 71
Napoleonic Wars 8, 12, 39, 47, 70
National Hospital for the Paralysed and
    Epileptic 148–9
nervous power 30, 75–6, 114

Newton, Sir Isaac 20–2, 28, 58
Noad, Henry Minchin 73, 120, 122
Normanby, Marquess of 9, 103–4, 107–8,
    110, 127

O'Key sisters 97
Ohm, Georg Simon 63, 130
    Ohm's law 160
Okes, Thomas Verney 18
Oxford 24, 132

Pall Mall Electric Association 138, 150–1, 153
Pall Mall Gazette 138, 143–4, 180–1
Pepper, John Henry 116–8, 121–3, 135, 141
    Pepper's ghost 117
Pepys, William Haseldine 26, 30
Perkins, Elisha 35–6, 38
Phipps, Constance 9, 102–11, 112–5, 117,
    121–3, 124, 126–7, 131, 134–5, 184
phrenology 96–8, 100
Pine, Thomas 77–8, 80
Poole, Thomas 64–5
Preece, William Henry 155, 167, 169
Priestley, Joseph 21, 27, 47, 64
Pulvermacher, J.L. 107–8, 151, 153
    Pulvermacher chain 107–8

Rennell, Thomas 45–6, 49, 54
resistance, human body 161–2, 166–7
Richardson, Benjamin Ward 118–9, 129,
    169–70
Ritter, Johan Wilhelm 28
Roget, Peter Mark 82, 87, 128
Royal College of Physicians 41, 146,
    154, 156–7, 162, 172, 182
Royal College of Surgeons 25, 40–1, 43–5,
    47, 55, 67, 105–6, 127, 149, 154
Royal Humane Society 26, 35
Royal Institution 25, 30–1, 34, 53, 55–6, 63,
    67–8, 70–2, 74, 92, 94–5, 98, 131, 134, 168
Royal Polytechnic Institution 72–4, 115–21,
    123, 124, 135, 141, 170
Royal Society 24, 30, 41, 43, 56, 61–4, 82,
    99, 128, 178
Rutter, John Obadiah Newell 86–7
Ryley, James Beresford 152–3

Salisbury, Marquess of 115
Sedgwick, Adam 65–7, 79
self–discipline see discipline
sex 34, 61, 88–9, 90, 127, 135, 152, 181

Sharples, Eliza 72, 78–80, 85
Shelley, Mary 37–8, 49
Shelley, Percy Bysshe 37
showmanship 8–9, 81, 91, 123, 135, 140–1, 143, 184
Smee, Alfred 67, 79–80
Society of Telegraph Engineers 146, 154–5, 161–2; *see also* Institute of Electrical Engineers
Somerville, Mary 58–9, 61–3, 92
Southey, Robert 33, 64
specialist hospitals 109–10, 119, 127, 145, 148–9
spectacle 9, 22, 55, 118, 184
spontaneous generation 40, 59
St Bartholomew's Hospital 34, 41, 43, 105, 146, 168
St Hilaire, Geoffroy 40
St Mary's Hospital 146, 161
St Thomas' Hospital 24, 34, 146–7, 154
Steavenson, William Edward 146–8, 156–7
Stewart, Balfour 130–3
Stone, William Henry 146, 154–5, 157, 161–2
Sturgeon, William 73, 76–7, 80, 81, 85–5

table–turning 96
Tait, Peter Guthrie 130
Tatum, Edward 164, 166
telegraph 9, 70–1, 73, 79, 82, 86, 106, 112–5, 118–9, 124, 126, 129–31, 133–4, 160
Tesla, Nikola 167–8
Thelwall, John 11, 27, 39
Thompson, Silvanus P. 155, 178
Thomson, William (Lord Kelvin) 130, 139, 144, 160, 178

Tibbits, Herbert 109–10, 126–7, 134, 148–50, 161, 176–8, 182
Tilt, Edward John 89
Tories 13, 19, 36, 40, 45–6, 48, 49, 103, 115
Tyndall, John 94–5, 131

Ure, Andrew 50–7, 169
uterine massage 154

vibrators 153–4
vital forces 34, 70, 128–9
vitalism 32–4, 38, 42–8, 79, 128–9
Volta, Alessandro 23–4, 29
vril 112–3, 123

Wakley, Thomas 41
Walker, Charles Vincent 73–4, 82, 86
Webber, Charles 154–5, 162
Weems, Mary Ann 12–4
Weems, Thomas 8, 11–9, 49–57, 70, 159, 184
West Riding Lunatic Asylum 134
West–end Hospital for Diseases of the Nervous System, Paralysis and Epilepsy 109, 149
Westinghouse, George 163–4, 166
Wheatstone, Charles 70, 115, 134
Whewell, William 65, 67, 74–5
Whigs 31, 61, 103–4
Wilkinson, Charles 29, 32–5, 37–8, 43, 136
Williams, Thomas 82
wonder, science of 8, 66, 72, 118, 123

Yatman, Matthew 32, 35–6

Zander, Gustav 138–40